Basics of DIFFERENTIAL CALCULUS

Dr. Sudhir Gupta

ANMOL PUBLICATIONS PVT. LTD.
NEW DELHI-110 002 (INDIA)

ANMOL PUBLICATIONS PVT. LTD.
Regd. Office: 4360/4, Ansari Road, Daryaganj,
New Delhi-110002 (India)
Tel.: 23278000, 23261597, 23286875, 23255577
Fax: 91-11-23280289
Email: anmolpub@gmail.com
Visit us at: www.anmolpublications.com

Branch Office: No. 1015, Ist Main Road, BSK IIIrd Stage
IIIrd Phase, IIIrd Block, Bengaluru-560 085 (India)
Tel.: 080-41723429 • Fax: 080-26723604
Email: anmolpublicationsbangalore@gmail.com

Basics of Differential Calculus

First Edition, 2011

ISBN 978-81-261-4828-8

PRINTED IN INDIA

Printed at Mehra Offset Press, Delhi.

Contents

Basics of
DIFFERENTIAL CALCULUS

Preface

Differential calculus is a subfield of calculus concerned with the study of the rates at which quantities change. It is one of the two traditional divisions of calculus, the other being integral calculus. The primary objects of study in differential calculus are the derivative of a function, related notions such as the differential, and their applications. The derivative of a function at a chosen input value describes the rate of change of the function near that input value.

The process of finding a derivative is called differentiation. Geometrically, the derivative at a point equals the slope of the tangent line to the graph of the function at that point. For a real-valued function of a single real variable, the derivative of a function at a point generally determines the best linear approximation to the function at that point.

Differential calculus and integral calculus are connected by the fundamental theorem of calculus, which states that differentiation is the reverse process to integration.

Author

Preface

Differential calculus is a subfield of calculus concerned with the study of the rates at which quantities change. It is one of the two traditional divisions of calculus, the other being integral calculus. The primary objects of study in differential calculus are the derivative of a function, related notions such as the differential, and their applications. The derivative of a function at a chosen input value describes the rate of change of the function near that input value.

The process of finding a derivative is called differentiation. Geometrically, the derivative at a point equals the slope of the tangent line to the graph of the function at that point. For a real-valued function of a single real variable, the derivative of a function at a point generally determines the best linear approximation to the function at that point.

Differential calculus and integral calculus are connected by the fundamental theorem of calculus, which states that differentiation is the reverse process to integration.

Author

1

Sequences of Functions

A sequence is a special kind of function whose domain is the positive integers. The range of a sequence is the collection of terms that make up the sequence. Just as the word sequence implies, the order of the terms in a sequence is important.

The first term of a sequence, for example, is found by taking the value of the function at 1; the second term is the value of the function at 2, and so on. Consider the sequence $f(x) = x$. The terms of the sequence, denoted $a_1, a_2, a_3, \ldots, a_n$ are $1, 2, 3, \ldots, n$. When working with sequences, instead of using function notation to express the formula of the function, a formula is the following form is used: $a_n = n$.

This is the same sequence as above, but the conventional n is used to denote an integer, since only integers are in the domain of sequences.

Two important categories of sequences are arithmetic sequences, and geometric sequences. Both are examples of a recursive sequence—a sequence in which each term (besides the first) depends on the previous term. Both of these types of sequences will be discussed.

When the terms of a sequence are summed, the result is called a series. Some series increase without bound as n increases, but others approach a limit. Both types of series will be studied in the following sections. There are also certain formulas for calculating the limits of series that we'll learn. The study of series is an important part of calculus, and it all starts with sequences.

SEQUENCES

Let's start off this section with a discussion of just what a sequence is. A sequence is nothing more than a list of numbers written in a specific order. The list may or may not have an infinite number of terms in them although we will be dealing exclusively with infinite sequences in this class. General sequence terms are denoted as follows,

$$\begin{aligned} &a_1 - \text{first term} \\ &a_2 - \text{second term} \\ &\vdots \\ &a_n - n^{th} \text{ term} \\ &a_{n+1} - (n+1)^{st} \text{ term} \\ &\vdots \end{aligned}$$

Because we will be dealing with infinite sequences each term in the sequence will be followed by another term as noted above. In the notation above we need to be very careful with the subscripts. The subscript of $n + 1$ denotes the next term in the sequence and NOT one plus the nth term! In other words,

$$a_{n+1} \neq a_n + 1$$

So be very careful when writing subscripts to make sure that the "+1" doesn't migrate out of the subscript! This is an easy mistake to make when you first start dealing with this kind of thing.

There is a variety of ways of denoting a sequence. Each of the following are equivalent ways of denoting a sequence.

$$\{a_1, a_2, \ldots, a_n, a_{n+1}, \ldots\} \qquad \{a_n\} \qquad \{a_n\}_{n+1}^{\infty}$$

In the second and third notations above a_n is usually given by a formula. A couple of notes are now in order about these notations. First, note the difference between the second and third notations above. If the starting point is not important or is implied in some way by the problem it is often not written down as we did in the third notation.

Next, we used a starting point of $n = 1$ in the third notation only so we could write one down. There is absolutely no reason to believe that a sequence will start at $n = 1$. A sequence will start where ever it needs to start.

Let's take a look at a couple of sequences.

Example: Write down the first few terms of each of the following sequences.

1. $\left\{\dfrac{n+1}{n^2}\right\}_{n=1}^{\infty}$

2. $\left\{\dfrac{(-1)^{n+1}}{2^n}\right\}_{n=0}^{\infty}$

3. $\{b_n\}_{n=0}^{\infty}$, where $b_n = n^{th}$ digit of π

Solution:

1. $\left\{\dfrac{n+1}{n^2}\right\}_{n=1}^{\infty}$

To get the first few sequence terms here all we need to do is plug in values of n into the formula given and we'll get the sequence terms.

$$\left\{\frac{n+1}{n^2}\right\}_{n=1}^{\infty} = \left\{\underbrace{2}_{n=1}, \underbrace{\frac{3}{4}}_{n=2}, \underbrace{\frac{4}{9}}_{n=3}, \underbrace{\frac{5}{16}}_{n=4}, \underbrace{\frac{6}{25}}_{n=5}, \ldots\right\}$$

Note the inclusion of the "..." at the end! This is an important piece of notation as it is the only thing that tells us that the sequence continues on and doesn't terminate at the last term.

2. $\left\{\dfrac{(-1)^{n+1}}{2^n}\right\}_{n=0}^{\infty}$

This one is similar to the first one. The main difference is that this sequence doesn't start at $n = 1$.

$$\left\{\frac{(-1)^{n+1}}{2^n}\right\}_{n=0}^{\infty} = \left\{-1, \frac{1}{2}, \frac{1}{4}, \frac{1}{8}, \frac{1}{16}, \ldots\right\}$$

Note that the terms in this sequence alternate in signs. Sequences of this kind are sometimes called alternating sequences.

3. $\{b_n\}_{n=1}^{\infty}$, where $b_n = n^{th}$ digit of π

This sequence is different from the first two in the sense that it doesn't have a specific formula for each term. However, it does tell us what each term should be. Each term should be the nth digit of π. So we know that p = 3.14159265359...

The sequence is then,

$$\{3, 1, 4, 1, 5, 9, 2, 6, 5, 3, 5,...\}$$

In the first two parts of the previous example note that we were really treating the formulas as functions that can only have integers plugged into them. Or,

$$f(n) = \frac{n+1}{n^2} \qquad g(n) = \frac{(-1)^{n+1}}{2^n}$$

This is an important idea in the study of sequences (and series). Treating the sequence terms as function evaluations will allow us to do many things with sequences that couldn't do otherwise.

Before delving further into this idea however we need to get a couple more ideas out of the way.

First we want to think about "graphing" a sequence. To graph the sequence $\{a_n\}$ we plot the points $(n_3\ a_n)$ as n ranges over all possible values on a graph. For instance, let's graph the sequence $\left\{\frac{n+1}{n^2}\right\}_{n=1}^{\infty}$. The first few points on the graph are,

$$(1, 2), \left(2, \frac{3}{3}\right), \left(3, \frac{4}{9}\right), \left(4, \frac{5}{16}\right), \left(5, \frac{6}{25}\right),...$$

The graph, for the first 30 terms of the sequence, is then,

This graph leads us to an important idea about sequences. Notice that as n increases the sequence terms from our sequence terms, in this case, get closer and closer to zero. We then say that zero is the limit (or sometimes the limiting value) of the sequence

and write,

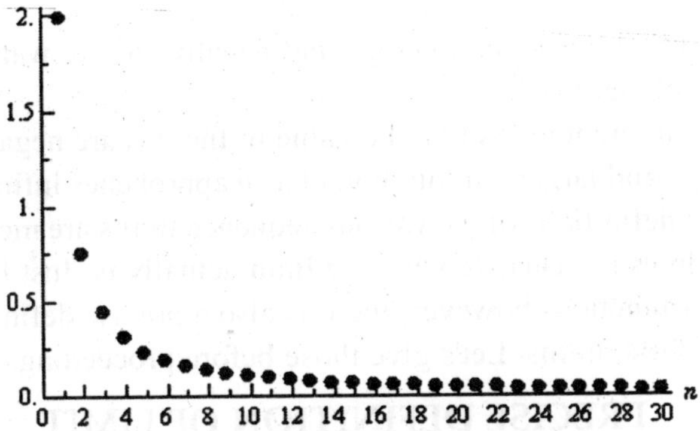

$$\lim_{n\to\infty} a_n = \lim_{n\to\infty} \frac{n+1}{n^2} = 0$$

This notation should look familiar to you. It is the same notation we used when we talked about the limit of a function. In fact, if you recall, we said earlier that we could think of sequences as functions in some way and so this notation shouldn't be too surprising. Using the ideas that we developed for limits of functions we can write down the following *working definition* for limits of sequences.

WORKING DEFINITION OF LIMIT

1. We say that

$$\lim_{n\to\infty} a_n = L$$

if we can make an as close to L as we want for all sufficiently large n. In other words, the value of the a_n's approach L as n approaches infinity.

2. We say that

$$\lim_{n\to\infty} a_n = \infty$$

if we can make a_n as large as we want for all sufficiently large n. Again, in other words, the value of the a_n's get larger and larger without bound as n approaches infinity.

3. We say that

$$\lim_{n\to\infty} a_n = -\infty$$

If we can make an as large and negative as we want for all sufficiently large n.

Again, in other words, the value of the a_n's are negative and get larger and larger without bound as n approaches infinity. The working definitions of the various sequence limits are nice in that they help us to visualize what the limit actually is. Just like with limits of functions however, there is also a precise definition for each of these limits. Let's give those before proceeding.

PRECISE DEFINITION OF LIMIT

1. We say that $\lim_{n\to\infty} a_n = L$ if for every number Ý > 0 there is an integer N such that
   $|an - L| <$ ⇑ whenever $n > N$
2. We say that $\lim_{n\to\infty} a_n = \infty$ if for every number $M > 0$ there is an integer N such that
   $an > M$ whenever $n > N$
3. We say that $\lim_{n\to\infty} a_n$ if for every number $M < 0$ there is an integer N such that
   $an < M$ whenever $n > N$

We won't be using the precise definition often, but it will show up occasionally.

Note that both definitions tell us that in order for a limit to exist and have a finite value all the sequence terms must be getting closer and closer to that finite value as n increases. Now that we have the definitions of the limit of sequences out of the way we have a bit of terminology that we need to look at.

If $\lim_{n\to\infty} a_n$ exists and is finite we say that the sequence is convergent. If $\lim_{n\to\infty} a_n$ doesn't exist or is infinite we say the sequence diverges. Note that sometimes we will say the sequence diverges to $-\infty$ if $\lim_{n\to\infty} a_n = \infty$ and if $\lim_{n\to\infty} a_n$ we will sometimes say that the sequence diverges to $-\infty$.

Get used to the terms "convergent" and "divergent" as we'll be seeing them quite a bit throughout this chapter.

So just how do we find the limits of sequences? Most limits of most sequences can be found using one of the following theorems.

Theorem: Given the sequence $\{a_n\}$ if we have a function $f(x)$ such that $f(n) = a_n$ an $\lim_{n\to\infty} f(x) = L$ then $\lim_{n\to\infty} a_n = L$

This theorem is basically telling us that we take the limits of sequences much like we take the limit of functions. In fact, in most cases we'll not even really use this theorem by explicitly writing down a function. We will more often just treat the limit as if it were a limit of a function and take the limit as we always did back in Calculus I when we were taking the limits of functions.

So, now that we know that taking the limit of a sequence is nearly identical to taking the limit of a function we also know that all the properties from the limits of functions will also hold.

PROPERTIES

1. $\lim_{n\to\infty} (a_n \pm b_n) = \lim_{n\to\infty} a_n \pm \lim_{n\to\infty} b_n$
2. $\lim_{n\to\infty} ca_n = c \lim_{n\to\infty} a_n$
3. $\lim_{n\to\infty} (a_n b_n) = \left(\lim_{n\to\infty} a_n\right)\left(\lim_{n\to\infty} b_n\right)$
4. $\lim_{n\to\infty} \frac{a_n}{b_n} = \frac{\lim_{n\to\infty} a_n}{\lim_{n\to\infty} b_n}$, provided $\lim_{n\to\infty} b_n \neq 0$
5. $\lim_{n\to\infty} a_n^p = \left[\lim_{n\to\infty} a_n\right]^p$ provided $a_n \geq 0$ provided

These properties can be proved using Theorem above and the function limit properties we saw in Calculus I or we can prove them directly using the precise definition of a limit using nearly identical proofs of the function limit properties. Next, just as we had a Squeeze Theorem for function limits we also have one for sequences and it is pretty much identical to the function limit version.

SQUEEZE THEOREM FOR SEQUENCES

If $a_n \le c_n \le b_n$ for all $n > N$ for some N and $\lim\limits_{n\to\infty} a_a \lim\limits_{n\to\infty} b_n = L$ then $\lim\limits_{n\to\infty} c_n = L$.

Note that in this theorem the "for all $n > N$ for some N" is really just telling us that we need to have $a_n \le c_n \le b_n$ for all sufficiently large n, but if it isn't true for the first few n that won't invalidate the theorem. As we'll see not all sequences can be written as functions that we can actually take the limit of. This will be especially true for sequences that alternate in signs. While we can always write these sequence terms as a function we simply don't know how to take the limit of a function like that. The following theorem will help with some of these sequences.

Theorem: If $\lim\limits_{n\to\infty} |a_n| = 0$ then $\lim\limits_{n\to\infty} a_n = 0$.

Note that in order for this theorem to hold the limit MUST be zero and it won't work for a sequence whose limit is not zero. This theorem is easy enough to prove so let's do that. Proof of Theorem: The main thing to this proof is to note that,

$$- \quad |a_n| \le a_n \le |a_n|$$

Then note that,

$$\lim_{n\to\infty} (-|a_n|) = -\lim_{n\to\infty} |a_n| = 0$$

We then have $\lim\limits_{n\to\infty} (-|a_n|) = \lim\limits_{n\to\infty} |a_n| = 0$ and so by the Squeeze Theorem we must also have,

$$\lim_{n\to\infty} a_n = 0$$

The next theorem is a useful theorem giving the convergence/ divergence and value (for when it's convergent) of a sequence that arises on occasion.

Theorem: The sequence $\left\{r^n\right\}_{n=0}^{\infty}$ converges if $-1 < r \le 1$ and diverges for all other value of r. Also,

$$\lim_{n\to\infty} r^n = \begin{cases} 0 & \text{if } -1 < r < 1 \\ 1 & \text{if } r = 1 \end{cases}$$

Here is a simple partial proof of this theorem.

Partial Proof of Theorem:

We'll do this by a series of cases although the last case will not be completely proven.

Case 1: $r > 1$

We know from Calculus I that $\lim_{n\to\infty} r^x = \infty$ if $r > 1$ and so by Theorem above we also know that $\lim_{n\to\infty} r^n = \infty$ and so the sequence diverges if $r > 1$.

Case 2: $r = 1$

In this case we have,

$$\lim_{n\to\infty} r^n = \lim_{n\to\infty} 1^n = \lim_{n\to\infty} 1 = 1$$

So, the sequence converges for $r = 1$ and in this case its limit is 1.

Case 3: $0 < r < 1$

We know from Calculus I that $\lim_{n\to\infty} r^x = 0$ if $0 < r < 1$ and so by Theorem above we also know that $\lim_{n\to\infty} r^n = 0$ and so the sequence converges if $0 < r < 1$ and in this case its limit is zero.

Case 4: $r = 0$

In this case we have,

$$\lim_{n\to\infty} r^n = \lim_{n\to\infty} 0^n = \lim_{n\to\infty} 0 = 0$$

So, the sequence converges for $r = 0$ and in this case its limit is zero.

Case 5: $-1 < r < 0$

First let's note that if $-1 < r < 0$ then $0 < |r| < 1$ then by Case 3 above we have,

$$\lim_{n\to\infty} |r^n| = \lim_{n\to\infty} |r^n| = 0$$

Theorem above now tells us that we must also have, $\lim_{n\to\infty} r^n = 0$ and so if $-1 < r < 0$ the sequence converges and has a limit of 0.

Case 6:

In this case the sequence is,

$$\left\{r^n\right\}_{n=0}^{\infty} = \left\{(-1)^n\right\}_{n=0}^{\infty} \qquad = \{1,-1,1,-1,1-1,1-1,...\}_{n=0}^{\infty}$$

and hopefully it is clear that $\lim_{n\to\infty} (-1)^n$ doesn't exist. Recall that in order of this limit to exist the terms must be approaching a single value as n increases. In this case however the terms just alternate between 1 and –1 and so the limit does not exist.

So, the sequence diverges for $r = -1$.

Case 7: $r < -1$

In this case we're not going to go through a complete proof. Let's just see what happens if we let $r = -2$ for instance. If we do that the sequence becomes,

$$\left\{r^n\right\}_{n=0}^{\infty} = \left\{(-2)^n\right\}_{n=0}^{\infty} = \{1, -2, 4, -8, 16, -32, ...\}_{n=0}^{\infty}$$

So, if $r = -2$ we get a sequence of terms whose values alternate in sign and get larger and larger and so $\lim_{n\to\infty} (-2)^n$ doesn't exist. It does not settle down to a single value as n increases nor do the terms ALL approach infinity. So, the sequence diverges for $r = -2$. We could do something similar for any value of r such that $r < -1$ and so the sequence diverges for $r < -1$.

Let's take a look at a couple of examples of limits of sequences.

Example: Determine if the following sequences converge or diverge. If the sequence converges determine its limit.

1. $\left\{\dfrac{3n^2-1}{10n+5n^2}\right\}_{n=2}^{\infty}$

2. $\left\{\dfrac{e^{2n}}{n}\right\}_{n=1}^{\infty}$

3. $\left\{\dfrac{(-1)^n}{n}\right\}_{n=1}^{\infty}$

4. $\left\{(-1)^n\right\}_{n=1}^{\infty}$

Solution:

1. $\left\{\dfrac{3n^2-1}{10n+5n^2}\right\}_{n=2}^{\infty}$

In this case all we need to do is recall the method that was developed in Calculus I to deal with the limits of rational functions

To do a limit in this form all we need to do is factor from the numerator and denominator the largest power of n, cancel and then take the limit.

$$\lim_{n\to\infty}\frac{3n^2-1}{10n+5n^2} = \lim_{n\to\infty}\frac{n^2\left(3-\frac{1}{n^2}\right)}{n^2\left(\frac{10}{n}+5\right)} = \lim_{n\to\infty}\frac{3-\frac{1}{n^2}}{\frac{10}{n}+5} = \frac{3}{5}$$

So the sequence converges and its limit is $\dfrac{3}{5}$.

2. $\left\{\dfrac{e^{2n}}{n}\right\}_{n=1}^{\infty}$

We will need to be careful with this one. We will need to use L'Hospital's Rule on this sequence. The problem is that L'Hospital's Rule only works on functions and not on sequences. Normally this would be a problem, but we've got Theorem from above to help us out. Let's define

$$f(x) = \frac{e^{2x}}{x}$$

and note that,

$$f(n) = \frac{e^{2n}}{n}$$

Theorem says that all we need to do is take the limit of the function.

$$\lim_{n\to\infty} \frac{e^{2n}}{n} = \lim_{n\to\infty} \frac{e^{2x}}{x} = \lim_{n\to\infty} \frac{2e^{2x}}{1} = \infty$$

So, the sequence in this part diverges (to ∞).

More often than not we just do L'Hospital's Rule on the sequence terms without first converting to *x's* since the work will be identical regardless of whether we use *x* or *n*.

However, we really should remember that technically we can't do the derivatives while dealing with sequence terms.

3. $\left\{\dfrac{(-1)^n}{n}\right\}_{n=1}^{\infty}$

We will also need to be careful with this sequence. We might be tempted to just say that the limit of the sequence terms is zero (and we'd be correct). However, technically we can't take the limit of sequences whose terms alternate in sign, because we don't know how to do limits of functions that exhibit that same behavior.

Also, we want to be very careful to not rely too much on

intuition with these problems. So, let's work this one by the book. We will need to use Theorem on this problem. To this we'll first need to compute,

$$\lim_{n\to\infty}\left|\frac{(-1)^n}{n}\right| = \lim_{n\to\infty}\frac{1}{n} = 0$$

Therefore, since the limit of the sequence terms with absolute value bars on them goes to zero we know by Theorem that,

$$\lim_{n\to\infty}\frac{(-1)^n}{n} = 0$$

Which also means that the sequence converges to a value of zero.

4. $\left\{(-1)^n\right\}_{n=0}^{\infty}$

For this theorem note that all we need to do is realize that this is the sequence in Theorem above using $r = -1$. So, by Theorem this sequence diverges. We now need to give a warning about misusing Theorem. Theorem only works if the limit is zero. If the limit of the absolute value of the sequence terms is not zero then the theorem will not hold. The last part of the previous example is a good example of this (and in fact this warning the whole reason that part is there). Notice that

$$\lim_{n\to\infty}\left|(-1)^n\right| = \lim_{n\to\infty} 1 = 1$$

and yet, $\lim_{n\to\infty}(-1)^n$ doesn't even exist let alone equal 1.So, be careful using this Theorem. Before moving onto the next section we need to give one more theorem that we'll need for a proof down the road.

Theorem: For the sequence $\{a_n\}$ if both $\lim_{n\to\infty} a_{2n} = L$ and $\lim_{n\to\infty} a_{2n+1} = L$ then $\{a_n\}$ is convergent and $\lim_{n\to\infty} a_n = L$.

Proof of Theorem:

Let å > 0.

Then since $\lim_{n\to\infty} a_{2n} = L$ there is an $N_1 > 0$ such that if $n > N_1$ we know that,

$$|a2n - L| \quad < å$$

Likewise, because,

$$\lim_{n\to\infty} a_{2n+1} = L$$

there is an $N_2 > 0$ such that if $n > N_2$ we know that,

$$|a_{2n+1} - L| \quad < å$$

Now, let $N = \max\{2N_1, 2N_2 + 1\}$ and let $n > N$. Then either $an = a_{2k}$ for some $k > N_1$ or $a_n = a_{2k+1}$ for some $k > N_2$ and so in either case we have that,

$$|a_n - L| < å$$

Therefore, $\lim_{n\to\infty} a_n = L$ and so $\{a_n\}$ is convergent.

MORE ON SEQUENCES

In the previous section we introduced the concept of a sequence and talked about limits of sequences and the idea of convergence and divergence for a sequence. In this section we want to take a quick look at some ideas involving sequences.

Let's start off with some terminology and definitions.

Given any sequence $\{a_n\}$ we have the following.

1. We call the sequence increasing if $a_n < a_{n+1}$ for every n.
2. We call the sequence decreasing if $a_n > a_{n+1}$ for every n.
3. If $\{a_n\}$ is an increasing sequence or $\{a_n\}$ is a decreasing sequence we call it monotonic.
4. If there exists a number m such that $m \le a_n$ for every n we say the sequence is bounded below. The number m is sometimes called a lower bound for the sequence.
5. If there exists a number M such that $a_n \le M$ for every n we say the sequence is bounded above. The number M is sometimes called an upper bound for the sequence.
6. If the sequence is both bounded below and bounded above we call the sequence bounded.

Note that in order for a sequence to be increasing or decreasing it must be increasing/decreasing for every n. In other words, a sequence that increases for three terms and then decreases for the rest of the terms is NOT a decreasing sequence!

Also note that a monotonic sequence must always increase or it must always decrease. Before moving on we should make a quick point about the bounds for a sequence that is bounded above and/or below. We'll make the point about lower bounds, but we could just as easily make it about upper bounds.

A sequence is bounded below if we can find any number m such that $m \leq a_n$ for every n. Note however that if we find one number m to use for a lower bound then any number smaller than m will also be a lower bound.

Also, just because we find one lower bound that doesn't mean there won't be a "better" lower bound for the sequence than the one we found. In other words, there are an infinite number of lower bounds for a sequence that is bounded below, some will be better than others.

In my class all that I'm after will be a lower bound. I don't necessarily need the best lower bound, just a number that will be a lower bound for the sequence.

Let's take a look at a couple of examples.

Example: Determine if the following sequences are monotonic and/or bounded.

1. $\left\{-n^2\right\}_{n=0}^{\infty}$

2. $\left\{(-1)^{n+1}\right\}_{n=1}^{\infty}$

3. $\left\{\dfrac{2}{n^2}\right\}_{n=5}^{\infty}$

Solution:

1. $\left\{-n^2\right\}_{n=0}^{\infty}$

This sequence is a decreasing sequence (and hence monotonic) because,

$$-n_2 \quad > -(n+1)^2$$

for every n.

Also, since the sequence terms will be either zero or negative this sequence is bounded above. We can use any positive number or zero as the bound, M, however, it's standard to choose the smallest possible bound if we can and it's a nice number. So, we'll choose $M = 0$ since,

$$-n_2 \quad \leq 0 \qquad \text{for every } n$$

This sequence is not bounded below however since we can always get below any potential bound by taking n large enough. Therefore, while the sequence is bounded above it is not bounded.

As a side note we can also note that this sequence diverges (to $-\infty$ if we want to be specific).

2. $\left\{(-1)^{n+1}\right\}_{n=1}^{\infty}$

The sequence terms in this sequence alternate between 1 and –1 and so the sequence is neither an increasing sequence or a decreasing sequence. Since the sequence is neither an increasing nor decreasing sequence it is not a monotonic sequence.

The sequence is bounded however since it is bounded above by 1and bounded below by -1. Again, we can note that this sequence is also divergent.

1. $\left\{\frac{2}{n^2}\right\}_{n=5}^{\infty}$

This sequence is a decreasing sequence (and hence monotonic) since,

$$\frac{2}{n^2} > \frac{2}{(n+1)^2}$$

The terms in this sequence are all positive and so it is bounded below by zero. Also, since the sequence is a decreasing sequence the first sequence term will be the largest and so we can see that the sequence will also be bounded above by $\frac{2}{25} \cdot \frac{2}{25}$ Therefore, this sequence is bounded. We can also take a quick limit and note that this sequence converges and its limit is zero.

Now, let's work a couple more examples that are designed to make sure that we don't get too used to relying on our intuition with these problems. As we noted in the previous section our intuition can often lead us astray with some of the concepts we'll be looking at in this chapter.

Example: Determine if the following sequences are monotonic and/or bounded.

1. $\left\{\dfrac{n}{n+1}\right\}_{n=1}^{\infty}$

2. $\left\{\dfrac{n^3}{n^4+10000}\right\}_{n=0}^{\infty}$

Solution:

1. $\left\{\dfrac{n}{n+1}\right\}_{n=1}^{\infty}$

We'll start with the bounded part of this example first and then come back and deal with the increasing/decreasing question since that is where students often make mistakes with this type of sequence. First, n is positive and so the sequence terms are all positive. The sequence is therefore bounded below by zero. Likewise each sequence term is the quotient of a number divided by a larger number and so is guaranteed to be less that one. The sequence is then bounded above by one. So, this sequence is bounded.

Now let's think about the monotonic question. First, students will often make the mistake of assuming that because the denominator is larger the quotient must be decreasing. This will not always be the case and in this case we would be wrong. This sequence is increasing as we'll see. To determine the increasing/ decreasing nature of this sequence we will need to resort to Calculus I techniques. First consider the following function and its derivative.

$$f(x) = \frac{x}{x+1} \qquad f(x)\frac{1}{(x+1)^2}$$

We can see that the first derivative is always positive and so from Calculus I we know that the function must then be an increasing function. So, how does this help us? Notice that,

$$f(n) = \frac{n}{n+1} = a_n$$

Therefore because $n < n + 1$ and $f(x)$ is increasing we can also say that,

$$a_n = \frac{n}{n+1} = f(n) < f(n+1) = \frac{n+1}{n+1+1} = a_{n+1}$$

In other words, the sequence must be increasing.

Note that now that we know the sequence is an increasing sequence we can get a better lower bound for the sequence. Since the sequence is increasing the first term in the sequence must be the smallest term and so since we are starting at $n = 1$ we could also use a lower bound of $\frac{1}{2}$ for this sequence. It is important to remember that any number that is always less than or equal to all the sequence terms can be a lower bound. Some are better than others however.

A quick limit will also tell us that this sequence converges with a limit of 1. Before moving on to the next part there is a natural question that many students will have at this point. Why did we use Calculus to determine the increasing/decreasing nature of the sequence when we could have just plugged in a couple of n's and quickly determined the same thing?

The answer to this question is the next part of this example!

2. $\left\{\dfrac{n^3}{n^4 + 10000}\right\}_{n=0}^{\infty}$

This is a messy looking sequence, but it needs to be in order to make the point of this part.

First, notice that, as with the previous part, the sequence terms are all positive and will all be less than one (since the numerator is guaranteed to be less than the denominator) and so the sequence is bounded. Now, let's move on to the increasing/decreasing question. As with the last problem, many students will look at the exponents in the numerator and denominator and determine based on that that sequence terms must decrease.

This however, isn't a decreasing sequence. Let's take a look at the first few terms to see this.

$$a_1 = \frac{1}{10001} \approx 0.0009999 \qquad a_2 = \frac{1}{1252} \approx 0.0007987$$

$$a_3 = \frac{27}{10081} \approx 0.005678 \qquad a_3 = \frac{4}{641} \approx 0.006240$$

$$a_5 = \frac{1}{85} \approx 0.011756 \qquad a_6 = \frac{27}{1412} \approx 0.019122$$

$$a_7 = \frac{343}{12401} \approx 0.02766 \qquad a_8 = \frac{32}{881} \approx 0.03632$$

$$a_9 = \frac{729}{16561} \approx 0.04402 \qquad a_{10} = \frac{1}{20} \approx 0.05$$

The first 10 terms of this sequence are all increasing and so clearly the sequence can't be a decreasing sequence. Recall that a sequence can only be decreasing if all the terms are decreasing. Now, we can't make another common mistake and assume that because the first few terms increase then whole sequence must also increase.

If we did that we would also be mistaken as this is also not an increasing sequence.

This sequence is neither decreasing or increasing. The only sure way to see this is to do the Calculus I approach to increasing/decreasing functions. In this case we'll need the following function and its derivative.

$$f(x) = \frac{x^3}{x^4 + 10000} \qquad f'(x)$$

$$= \frac{-x^2(x^4 - 30000)}{(x^4 + 10000)^2}$$

This function will have the following three critical points,

$x = 0, x = \sqrt[4]{30000} \approx 13.1607, \qquad x = -\sqrt[4]{30000} \approx -13.1607$

Why critical points? Remember these are the only places where the function may change sign! Our sequence starts at $n = 0$ and so we can ignore the third one since it lies outside the values of n that we're considering.

By plugging in some test values of x we can quickly determine that the derivative is positive for $0 < x < \sqrt[4]{30000} \approx 13.16$ and so the function is increasing in this range. Likewise, we can see that the derivative is negative for $x > \sqrt[4]{30000} \approx 13.16$ and so the function will be decreasing in this range.

So, our sequence will be increasing for $0 \leq n \leq 13$ and decreasing for $n \geq 13$.

Therefore the function is not monotonic. Finally, note that this sequence will also converge and has a limit of zero. So, as the last example has shown we need to be careful in making assumptions about sequences.

Our intuition will often not be sufficient to get the correct answer and we can NEVER make assumptions about a sequence based on the value of the first few terms. As the last part has shown there are sequences which will increase or decrease for a few terms and then change direction after that.

Note as well that we said "first few terms" here, but it is completely possible for a sequence to decrease for the first 10,000 terms and then start increasing for the remaining terms. In other words, there is no "magical" value of n for which all we have to do is check up to that point and then we'll now that the whole sequence will do.

The only time that we'll be able to avoid using Calculus techniques to determine the increasing/decreasing nature of a sequence is in sequences like part (c) of Example 1. In this case increasing n only changed (in fact increased) the denominator and so we were able to determine the behavior of the sequence based on that.

In Example however, increasing n inc:eased both the denominator and the numerator. In cases like this there is no way to determine which increase will "win out" and cause the sequence terms to increase or decrease and so we need to resort to Calculus techniques to answer the question. We'll close out this section with a nice theorem that we'll use in some of the proofs later in this chapter.

Theorem: If $\{a_n\}$ is bounded and monotonic then $\{a_n\}$ is convergent.

Be careful to not misuse this theorem. It does not say that if a sequence is not bounded and/or not monotonic that it is divergent. Example 2*b* is a good case in point. The sequence in that example was not monotonic but it does converge.

Note as well that we can make several variants of this theorem. If $\{a_n\}$ is bounded above and increasing then it converges and likewise if $\{a_n\}$ is bounded below and decreasing then it converges.

2

Basics of Series

In this section we will introduce the topic that we will be discussing for the rest of this chapter. That topic is infinite series. So just what is a infinite series? Well, let's start with a sequence $\{a_n\}_{n=1}^{\infty}$ (note the $n = 1$ is for convenience, it can be anything) and define the following,

$$\begin{aligned} s_1 &= a_1 \\ s_2 &= a_1 + a_2 \\ s_3 &= a_1 + a_2 + a_3 \\ s_4 &= a_1 + a_2 + a_3 + a_4 \\ &\vdots \\ s_n &= a_1 + a_2 + a_3 + a_4 + \ldots + a_n = \sum_{i=1}^{n} a_i \end{aligned}$$

The are called partial sums and notice that they will form a sequence, $\{s_n\}_{n=1}^{\infty}$. Also recall that the Σ is used to represent this summation and called a variety of names.

The most common names are: series notation, summation notation, and sigma notation.

Now back to series. We want to take a look at the limit of the sequence of partial sums, $\{s_n\}_{n=1}^{\infty}$. Notationally we'll define,

$$\lim_{n\to\infty} s_n = \lim_{n\to\infty} \sum_{i=1}^{n} a_i = \sum_{i=1}^{\infty} a_i$$

We will call $\sum_{i=1}^{\infty} a_i$ an infinite series and note that the series "starts" at $i = 1$ because that is where our original sequence, $\{s_n\}_{n=1}^{\infty}$, started. Had our original sequence started at 2 then our infinite series would also have started at 2. The infinite series will start at the same value that the sequence of terms (as opposed to the sequence of partial sums) starts.

If the sequence of partial sums, $\{s_n\}_{n=1}^{\infty}$, is convergent and its limit is finite then we also call the infinite series, $\sum_{i=1}^{\infty} a_i$ convergent and if the sequence of partial sums is diverent then the infinite series is also called divergent. Note that sometimes it is convenient to write the infinite series as,

$$\sum_{i=1}^{\infty} a_i = a_1 + a_2 + a_3 + ... + a_n + ...$$

We do have to be careful with this however. This implies that an infinite series is just an infinite sum of terms and as well see in the next section this is not really true.

In the next section we're going to be discussing in greater detail the value of an infinite series, provided it has one of course as well as the ideas of convergence and divergence. In this section is going to be devoted mostly to notational issues as well as making sure we can do some basic manipulations with infinite series so we are ready for them when we need to be able to deal with them in later sections.

First, we should note that in most of this chapter we will refer to infinite series as simply series. If we ever need to work with both infinite and finite series we'll be more careful with terminology, but in most sections we'll be dealing exclusively with infinite series and so we'll just call them series.

Now, in $\sum_{i=1}^{\infty} a_i$ the i is called the index of summation or just index for short and note that the letter we use to represent the index does not matter. So for example the following series are all the same. The only difference is the letter we've used for the index.

$$\sum_{i=0}^{\infty}\frac{3}{i^2+1}=\sum_{i=0}^{\infty}\frac{3}{k^2+1}=\sum_{i=0}^{\infty}\frac{3}{n^2+1}\ etc.$$

It is important to again note that the index will start at whatever value the sequence of series terms starts at and this can literally be anything. So far we've used $n = 0$ and $n = 1$ but the index could have started anywhere. In fact, we will usually use $\sum a_n$ to represent an infinite series in which the starting point for the index is not important. When we drop the initial value of the index we'll also drop the infinity from the top so don't forget that it is still technically there. We will be dropping the initial value of the index in quite a few facts and theorems that we'll be seeing throughout this chapter. In these facts/theorems the starting point of the series will not affect the result and so to simplify the notation and to avoid giving the impression that the starting point is important we will drop the index from the notation. Do not forget however, that there is a starting point and that this will be an infinite series. Note however, that if we do put an initial value of the index on a series in a fact/theorem it is there because it really does need to be there. Now that some of the notational issues are out of the way we need to start thinking about various ways that we can manipulate series. We'll start this off with basic arithmetic with infinite series as we'll need to be able to do that on occasion. We have the following properties.

PROPERTIES

If $\sum a_n$ and $\sum b_n$ are both convergent series then,

1. $\sum ca_n$, where c is any number, is also convergent and

$$\sum ca_n = c\sum a_n$$

2. $\sum_{n=k}^{\infty} a_n \pm \sum_{n=k}^{\infty} b_n$ is also convergent and,

$$\sum_{n=k}^{\infty} a_n \pm \sum_{n=k}^{\infty} b_n = \sum_{n=k}^{\infty} (a_n \pm b_n).$$

The first property is simply telling us that we can always factor a multiplicative constant out of an infinite series and again recall that if we don't put in an initial value of the index that the series can start at any value. Also recall that in these cases we won't put an infinity at the top either.

The second property says that if we add/subtract series all we really need to do is add/subtract the series terms. Note as well that in order to add/subtract series we need to make sure that both have the same initial value of the index and the new series will also start at this value. Before we move on to a different topic let's discuss multiplication of series briefly. We'll start both series at $n = 0$ for a later formula and then note that,

$$\left(\sum_{n=0}^{\infty} a_n\right)\left(\sum_{n=0}^{\infty} b_n\right) \neq \sum_{n=0}^{\infty} (a_n b_n)$$

To convince yourself that this isn't true consider the following product of two finite sums.

$$(2 + x)(3 - 5x + x^2) = 6 - 7x - 3x^2 + x^3$$

Yeah, it was just the multiplication of two polynomials. Each is a finite sum and so it makes the point. In doing the multiplication we didn't just multiply the constant terms, then the x terms, *etc.* Instead we had to distribute the 2 through the second polynomial, then distribute the x through the second polynomial and finally combine like terms.

Multiplying infinite series (even though we said we can't think of an infinite series as an infinite sum) needs to be done in the same manner. With multiplication we're really asking us to do the following,

$$\left(\sum_{n=0}^{\infty} a_n\right)\left(\sum_{n=0}^{\infty} b_n\right) = (b_0 + b_1 + b_2 + b_3 + \dots)$$

To do this multiplication we would have to distribute the a_0 through the second term, distribute the a_1 through, etc then combine like terms. This is pretty much impossible since both series have an infinite set of terms in them, however the following formula can be used to determine the product of two series.

$$\left(\sum_{n=0}^{\infty} a_n\right)\left(\sum_{n=0}^{\infty} b_n\right) = \sum_{n=0}^{\infty} c_n \text{ where } cn = \sum_{n=0}^{\infty} a_i b_{n+1}$$

We also can't say a lot about the convergence of the product. Even if both of the original series are convergent it is possible for the product to be divergent.

The reality is that multiplication of series is a somewhat difficult process and in general is avoided if possible. We will take a brief look at it towards the end of the chapter when we've got more work under our belt and we run across a situation where it might actually be what we want to do. Until then, don't worry about multiplying series.

The next topic that we need to discuss in this section is that of index shift. To be honest this is not a topic that we'll see all that often in this course. In fact, we'll use it once in the next section and then not use it again in all likelihood.

Despite the fact that we won't use it much in this course doesn't mean however that it isn't used often in other classes where you might run across series. So, we will cover it briefly here so that you can say you've seen it. The basic idea behind index shifts is to start a series at a different value for whatever the reason (and yes, there are legitimate reasons for doing that).

Consider the following series,

$$\sum_{i=2}^{\infty} \frac{n+5}{2^n}$$

Suppose that for some reason we wanted to start this series at $n = 0$, but we didn't want to change the value of the series. This means that we can't just change the $n = 2$ to $n = 0$ as this would add in two new terms to the series and thus changing its value. Performing an index shift is a fairly simple process to do.

We'll start by defining a new index, say i, as follows,

$$i = n - 2$$

Now, when $n = 2$, we will get $i = 0$. Notice as well that if $n = \infty$ then $i = \infty - 2$, so only the lower limit will change here. Next, we can solve this for n to get,

$$n = i + 2$$

We can now completely rewrite the series in terms of the index i instead of the index n simply be plugging in our equation for n in terms of i.

$$\sum_{i=2}^{\infty} \frac{n+5}{2^n} = \sum_{i=0}^{\infty} \frac{(i+2)+5}{2^{i+2}} = \sum_{i=0}^{\infty} \frac{i+7}{2^{i+2}}$$

To finish the problem out we'll recall that the letter we used for the index doesn't matter and so we'll change the final i back into an n to get,

$$\sum_{i=2}^{\infty} \frac{n+5}{2^n} = \sum_{i=0}^{\infty} \frac{n+7}{2^{n+2}}$$

To convince yourselves that these really are the same summation let's write out the first couple of terms for each of them,

$$\sum_{i=2}^{\infty} \frac{n+5}{2^n} = \frac{7}{2^2} + \frac{8}{2^3} + \frac{9}{2^4} + \frac{10}{2^5} + \ldots$$

$$\sum_{i=2}^{\infty} \frac{n+7}{2^{n+2}} = \frac{7}{2^2} + \frac{8}{2^3} + \frac{9}{2^4} + \frac{10}{2^5} + \ldots$$

So, sure enough the two series do have exactly the same terms.

There is actually an easier way to do an index shift. The method given above is the technically correct way of doing an index shift. However, notice in the above example we decreased the initial value of the index by 2 and all the n's in the series terms increased by 2 as well. This will always work in this manner.

If we decrease the initial value of the index by a set amount then all the other n's in the series term will increase by the same amount. Likewise, if we increase the initial value of the index by a set amount, then all the n's in the series term will decrease by the same amount. Let's do a couple of examples using this shorthand method for doing index shifts.

Example: Perform the following index shifts.

1. Write $\sum_{i=1}^{\infty} ar^{n-1}$ as a series that starts at $n = 0$.

2. Write $\sum_{i=1}^{\infty} \frac{n^2}{1-3^{n+1}}$ as a series that starts at $n = 3$.

Solution:

1. In this case we need to decrease the initial value by 1 and so the n's (okay the single n) in the term must increase by 1 as well.

$$\sum_{i=1}^{\infty} ar^{n-1} = \sum_{i=1}^{\infty} ar^{(n+1)-1} = \sum_{i=1}^{\infty} ar^{n}$$

2. For this problem we want to increase the initial value by 2 and so all the n's in the series term must decrease by 2.

$$\sum_{i=1}^{\infty} \frac{n^2}{1-3^{n+1}} = \sum_{i=3}^{\infty} \frac{(n-2)^2}{1-3^{(n-2)+1}} = \sum_{i=3}^{\infty} \frac{(n-2)^2}{1-3^{n-1}}$$

The final topic in this section is again a topic that we'll not be seeing all that often in this class, although we will be seeing it more often than the index shifts. This final topic is really more about alternate ways to write series when the situation requires it.

Let's start with the following series and note that the $n = 1$ starting point is only for convenience since we need to start the series somewhere.

$$\sum_{n=1}^{\infty} a_n = a_1 + a_2 + a_3 + a_4 + a_5 + ...$$

Notice that if we ignore the first term the remaining terms will also be a series that will start at $n = 2$ instead of $n = 1$ So, we can rewrite the original series as follows,

$$\sum_{n=1}^{\infty} a_n = a_1 + \sum_{n=2}^{\infty} a_n$$

In this example we say that we've stripped out the first term. We could have stripped out more terms if we wanted to. In the following series we've stripped out the first two terms and the first four terms respectively.

$$\sum_{n=1}^{\infty} a_n = a_1 + a_2 + \sum_{n=3}^{\infty} a_n$$

$$\sum_{n=1}^{\infty} a_n = a_1 + a_2 + a_3 + a_4 + \sum_{n=5}^{\infty} a_n$$

Being able to strip out terms will, on occasion, simplify our work or allow us to reuse a prior result so it's an important idea to remember. Notice that in the second example above we could have also denoted the four terms that we stripped out as a finite series as follows,

$$\sum_{n=1}^{\infty} a_n = a_1 + a_2 + a_3 + a_4 + \sum_{n=5}^{\infty} a_n = \sum_{n=1}^{4} a_n + \sum_{n=5}^{\infty} a_n$$

This is a convenient notation when we are stripping out a large number of terms or if we need to strip out an undetermined number of terms. In general, we can write a series as follows,

$$\sum_{n=1}^{\infty} a_n = \sum_{n=1}^{N} a_n + \sum_{n=N+1}^{\infty} a_n$$

We'll leave this section with an important warning about terminology. Don't get sequences and series confused! A sequence is a list of numbers written in a specific order while an infinite series is a limit of a sequence of finite series and hence, if it exists will be a single value. So, once again, a sequence is a list of numbers while a series is a single number, provided it makes sense to even compute the series. Students will often confuse the two and try to use facts pertaining to one on the other. However, since they are different beasts this just won't work. There will be problems where we are using both sequences and series so we'll always have to remember that they are different.

CONVERGENCE/DIVERGENCE

In the previous section we spent some time getting familiar with series and we briefly defined convergence and divergence. Before worrying about convergence and divergence of a series we wanted to make sure that we've started to get comfortable with the notation involved in series and some of the some of the various manipulations of series that we will, on occasion, need to be able to do. As noted in the previous section most of what we were doing there won't be done much in this chapter.

So, it is now time to start talking about the convergence and divergence of a series as this will a topic that we'll be dealing with to one extent of another in almost all of the remaining sections of this chapter. So, let's recap just what an infinite series is and what it means for a series to be convergent or divergent.

We'll start with a sequence $\{a_n\}_{n=1}^{\infty}$ and again note that we're starting the sequence at $n = 1$ only for the sake of convenience and it can, in fact, be anything.

Next we define the partial sums of the series as,

$$\begin{aligned}
s_1 &= a_1 \\
s_2 &= a_1 + a_2 \\
s_3 &= a_1 + a_2 + a_3 \\
s_4 &= a_1 + a_2 + a_3 + a_4 \\
&\vdots \\
s_n &= a_1 + a_2 + a_3 + a_4 + \ldots + a_n = \sum_{i=1}^{n} a_i
\end{aligned}$$

and these form a new sequence, $\{s_n\}_{n=1}^{\infty}$.

An infinite series, or just series here since almost every series that will be looking at will be an infinite series, is then the limit of the partial sums. Or,

$$\sum_{i=1}^{\infty} a_i = \lim_{n \to \infty} s_n$$

If the sequence of partial sums is a convergent sequence

(*i.e.* its limit exists and is finite) then the series is also called convergent and in this case if $\lim_{n\to\infty} s_n = s$ then, $\sum_{i=1}^{\infty} a_i = s$. Likewise, if the sequence of partial sums is a divergent sequence (i.e. its limit doesn't exist or is plus or minus infinity) then the series is also called divergent.

Let's take a look at some series and see if we can determine if they are convergent or divergent and see if we can determine the value of any convergent series we find.

Example: Determine if the following series is convergent or divergent. If it converges determine its value.

$$\sum_{i=1}^{\infty} n$$

Solution: To determine if the series is convergent we first need to get our hands on a formula for the general term in the sequence of partial sums.

$$s_n = \sum_{i=1}^{n} i$$

This is a known series and its value can be shown to be,

$$sn = \sum_{i=1}^{n} i = \frac{n(n+1)}{2}$$

Don't worry if you didn't know this formula (I'd be surprised if anyone knew it...) as you won't be required to know it in my course. So, to determine if the series is convergent we will first need to see if the sequence of partial sums,

$$\left\{\frac{n(n+1)}{2}\right\}_{n=4}^{\infty}$$

Is convergent or divergent. That's not terribly difficult in this case. The limit of the sequence terms is,

$$\lim_{n\to\infty} \frac{n(n+1)}{2} = \infty$$

Therefore, the sequence of partial sums diverges to and so the series also diverges.

So, as we saw in this example we had to know a fairly obscure formula in order to determine the convergence of this series. In general finding a formula for the general term in the sequence of partial sums is a very difficult process.

In fact after the next section we'll not be doing much with the partial sums of series due to the extreme difficulty faced in finding the general formula. This also means that we'll not be doing much work with the value of series since in order to get the value we'll also need to know the general formula for the partial sums.

We will continue with a few more examples however, since this is technically how we determine convergence and the value of a series. Also, the remaining examples we'll be looking at in this section will lead us to a very important fact about the convergence of series. So, let's take a look at a couple more examples.

Example: Determine if the following series converges or diverges. If it converges determine its sum.

$$\sum_{n=2}^{\infty} \frac{1}{n^2 - 1}$$

Solution: This is actually one of the few series in which we are able to determine a formula for the general term in the sequence of partial fractions. However, in this section we are more interested in the general idea of convergence and divergence and so we'll put off discussing the process for finding the formula until the next section. The general formula for the partial sums is,

$$s_n = \sum_{i=2}^{n} \frac{1}{i^2 - 1} = \frac{3}{4} - \frac{1}{2n} - \frac{1}{2(n+1)}$$

and in this case we have,

$$\lim_{n\to\infty} s_n = \lim_{n\to\infty} \left(\frac{3}{4} - \frac{1}{2n} - \frac{1}{2(n+1)} \right) = \frac{3}{4}$$

The sequence of partial sums converges and so the series converges also and its value is,

$$\sum_{i=2}^{n} \frac{1}{n^2 - 1} = \frac{3}{4}$$

Example: Determine if the following series converges or diverges. If it converges determine its sum.

$$\sum_{n=0}^{\infty} (-1)^n$$

Solution: In this case we really don't need a general formula for the partial sums to determine the convergence of this series. Let's just write down the first few partial sums.

$$s_0 = 1$$
$$s_1 = 1 - 1 = 0$$
$$s_2 = 1 - 1 + 1 = 1$$
$$s_3 = 1 - 1 + 1 - 1 = 0 \quad \textit{etc.}$$

So, it looks like the sequence of partial sums is,

$$\{s_n\}_{n=0}^{\infty} = \{1, 0, 1, 0, 1, 0, 1, 0, 1,...\}$$

and this sequence diverges since $\lim_{n\to\infty} s_n$ doesn't exist. Therefore, the series also diverges.

Example: Determine if the following series converges or diverges. If it converges determine its sum.

$$\sum_{n=1}^{\infty} \frac{1}{3^{n-1}}$$

Solution: Here is the general formula for the partial sums for this series.

$$sn = \sum_{i=1}^{n} \frac{1}{3^{i-1}} = \frac{3}{2}\left(1 - \frac{1}{3^n}\right)$$

Again, do not worry about knowing this formula. This is not something that you'll ever be asked to know in my class.

In this case the limit of the sequence of partial sums is,

$$\lim_{n\to\infty} s_n = \lim_{n\to\infty} \frac{3}{2}\left(1 - \frac{1}{3^n}\right) = \frac{3}{2}$$

The sequence of partial sums is convergent and so the series will also be convergent. The value of the series is,

$$\sum_{i=1}^{\infty}\frac{1}{3^{n-1}}=\frac{3}{2}$$

As we already noted, do not get excited about determining the general formula for the sequence of partial sums. There is only going to be one type of series where you will need to determine this formula and the process in that case isn't too bad. In fact, you already know how to do most of the work in the process as you'll see in the next section.

So, we've determined the convergence of four series now. Two of the series converged and two diverged. Let's go back and examine the series terms for each of these. For each of the series let's take the limit as n goes to infinity of the series terms (not the partial sums!!).

$\lim_{n\to\infty} n=\infty$ this series diverged

$\lim_{n\to\infty} \frac{1}{n^2-1}=0$ this series converged

$\lim_{n\to\infty} (-1)^n$ doesn t exist this series diverged

$\lim_{n\to\infty} \frac{1}{3^{2-1}}=0$ this series converged

Notice that for the two series that converged the series term itself was zero in the limit. This will always be true for convergent series and leads to the following theorem.

Theorem:

If $\sum a_n$ converges then $\lim_{n\to\infty} a_n=0$.

Proof: First let's suppose that the series starts at $n = 1$. If it doesn't then we can modify things as appropriate below. Then the partial sums are,

$$s_{n-1}=\sum_{i=1}^{n-1}a_i=a_1+a_2+a_3+a_4+...+a_{n-1}$$

$$s_n=\sum_{i=1}^{n}a_i=a_1+a_2+a_3+a_4+...+a_{n-1}+a_n$$

Next, we can use these two partial sums to write,

$$a_n = s_n - s_{n-4}$$

Now because we know that $\sum a_n$ is convergent we also know that the sequence $\{s_n\}_{n=1}^{\infty}$ is also convergent and that $\lim_{n\to\infty} s_n = s$ for some finite value s. However, since $n - 1 \to \infty$ as $n \to \infty$ $n \to \infty$ we also have $\lim_{n\to\infty} s_{n-1} = s$.

We now have,

$$\lim_{n\to\infty} a_n = \lim_{n\to\infty} (s_n - s_{n-1}) = \lim_{n\to\infty} s_n - \lim_{n\to\infty} s_{n-1} = s - s = 0$$

Be careful to not misuse this theorem! This theorem gives us a requirement for convergence but not a guarantee of convergence. In other words, the converse is NOT true. If $\lim_{n\to\infty} a_n = 0$ the series may actually diverge! Consider the following two series.

$$\sum_{n=1}^{\infty} \frac{1}{n} \qquad \sum_{n=1}^{\infty} \frac{1}{n^2}$$

In both cases the series terms are zero in the limit as n goes to infinity, yet only the second series converges. The first series diverges.

It will be a couple of sections before we can prove this, so at this point please believe this and know that you'll be able to prove the convergence of these two series in a couple of sections.

Again, as noted above, all this theorem does is give us a requirement for a series to converge. In order for a series to converge the series terms must go to zero in the limit. If the series terms do not go to zero in the limit then there is no way the series can converge since this would violate the theorem.

DIVERGENCE TEST

If $\lim_{n\to\infty} a_n \neq 0$ then $\sum a_n$ will diverge.

Again, do NOT misuse this test. This test only says that a

series is guaranteed to diverge if the series terms don't go to zero in the limit. If the series terms do happen to go to zero the series may or may not converge! Again, recall the following two series,

$$\sum_{n=1}^{\infty} \frac{1}{n} \text{ diverges}$$

$$\sum_{n=1}^{\infty} \frac{1}{n^2} \text{ converges}$$

One of the more common mistakes that students make when the first get into series is to assume that if $\lim_{n \to \infty} a_n = 0$ $\lim_{n \to \infty} a_n = 0$ then $\sum a_n$ will converge. There is just no way to guarantee this so be careful!

Example: Determine if the following series is convergent or divergent.

$$\sum_{n=0}^{\infty} \frac{4n^2 - n^3}{10 + 2n^3}$$

Solution: With almost every series we'll be looking at in this chapter the first thing that we should do is take a look at the series terms and see if they go to zero of not. If it's clear that the terms don't go to zero use the Divergence Test and be done with the problem.

That's what we'll do here.

$$\lim_{n \to \infty} \frac{4n^2 - n^3}{10 + 2n^3} = -\frac{1}{2} \neq 0$$

The limit of the series terms isn't zero and so by the Divergence Test the series diverges.

The divergence test is the first test of many tests that we will be looking at over the course of the next several sections. You will need to keep track of all these tests, the conditions under which they can be used and their conclusions all in one place so you can quickly refer back to them as you need to.

Next we should talk briefly revisit arithmetic of series and convergence/divergence. As we saw in the previous section if $\sum a_n$ and $\sum b_n$ are both convergent series then so are $\sum ca_n$ and $\sum_{n=k}^{\infty}(a_n \pm b_n)$. Furthermore, these series will have the following sums or values.

$$\sum ca_n = c\sum a_n \quad \sum_{n=k}^{\infty}(a_n \pm b_n) = \sum_{n=k}^{\infty} a_n \pm \sum_{n=k}^{\infty} b_n$$

We'll see an example of this in the next section after we get a few more examples under our belt. At this point just remember that a sum of convergent sequences is convergent and multiplying a convergent sequence by a number will not change its convergence. We need to be a little careful with these facts when it comes to divergent series.

In the first case if $\sum a_n$ is divergent then $\sum ca_n$ will also be divergent (provided c isn't zero of course) since multiplying a series that is infinite in value or doesn't have a value by a finite value (*i.e.* c) won't change the fact that the series has an infinite or no value. However, it is possible to have both $\sum a_n$ and $\sum b_n$ be divergent series and yet have,

$$\sum_{n=k}^{\infty}(a_n \pm b_n)$$

be a convergent series. Now, since the main topic of this section is the convergence of a series we should mention a stronger type of convergence.

A series $\sum a_n$ is said to converge absolutely if $\sum |a_n|$ also converges. Absolute convergence is stronger than convergence in the sense that a series that is absolutely convergent will also be convergent, but a series that is convergent may or may not be absolutely convergent.

In fact if $\sum a_n$ converges and $\sum |a_n|$ diverges the series $\sum a_n$ is called conditionally convergent.

At this point we don't really have the tools at hand to properly investigate this topic in detail nor do we have the tools in hand to determine if a series is absolutely convergent or not. So we'll not say anything more about this subject for a while. When we finally have the tools in hand to discuss this topic in more detail we will revisit it. Until then don't worry about it. The idea is mentioned here only because we were already discussing convergence in this section and it ties into the last topic that we want to discuss in this section.

In the previous section after we'd introduced the idea of an infinite series we commented on the fact that we shouldn't think of an infinite series as an infinite sum despite the fact that the notation we use for infinite series seems to imply that it is an infinite sum. It's now time to briefly discuss this. First, we need to introduce the idea of a rearrangement. A rearrangement of a series is exactly what it might sound like, it is the same series with the terms rearranged into a different order.

For example, consider the following the infinite series.

$$\sum_{n=1}^{\infty} a_n = a_1 + a_2 + a_3 + a_4 + a_5 + a_6 + a_7 + ...$$

A rearrangement of this series is,

$$\sum_{n=1}^{\infty} a_n = a_2 + a_1 + a_3 + a_{14} + a_5 + a_9 + a_4 + ...$$

The issue we need to discuss here is that for some series each of these arrangements of terms can have a different values despite the fact that they are using exactly the same terms.

Here is an example of this. It can be shown that,

$$\sum_{n=1}^{\infty} \frac{(-1)^{n+1}}{n} = 1 - \frac{1}{2} + \frac{1}{3} - \frac{1}{4} + \frac{1}{5} - \frac{1}{6} + \frac{1}{7} - \frac{1}{8} + ... = \text{In } 2$$

Since this series converges we know that if we multiply it by a constant c its value will also be multiplied by c. So, let's multiply this by,

$$\frac{1}{2}$$

to get,

$$\frac{1}{2}-\frac{1}{4}+\frac{1}{6}-\frac{1}{8}+\frac{1}{10}-\frac{1}{12}+\frac{1}{14}+\frac{1}{16}...=\frac{1}{2} \quad = \text{In } 2$$

Now, let's add in a zero between each term as follows.

$$0+\frac{1}{2}+0-\frac{1}{4}+0+\frac{1}{6}+0-\frac{1}{8}+0+\frac{1}{10}+0\frac{1}{12}+0+...=\frac{1}{2} \text{ In } 2$$

We know that if two series converge we can add them by adding term by term

$$1+\frac{1}{3}-\frac{1}{2}+\frac{1}{5}+\frac{1}{7}-\frac{1}{4}+...=\frac{3}{2} \text{ In } 2$$

FACTS

Given the series $\sum a_n$,

1. If $\sum a_n$ is absolutely convergent and its value is s then any rearrangement of $\sum a_n$ will also have a value of s.
2. If $\sum a_n$ is conditionally convergent and r is any real number then there is a rearrangement of $\sum a_n$ whose value will be r.

Again, we do not have the tools in hand yet to determine if a series is absolutely convergent and so don't worry about this at this point. This is here just to make sure that you understand that we have to be very careful in thinking of an infinite series as an infinite sum. There are times when we can (i.e. the series is absolutely convergent) and there are times when we can't (i.e. the series is conditionally convergent).

As a final note, the fact above tells us that the series,

$$\sum_{n=1}^{\infty}\frac{(-1)^{n+1}}{n}$$

must be conditionally convergent since two rearrangements gave two separate values of this series. Eventually it will be very simple to show that this series conditionally convergent.

SPECIAL SERIES

In this section we are going to take a brief look at three special series. Actually, special may not be the correct term. All three have been named which makes them special in some way, however the main reason that we're going to look at two of them in this section is that they are the only types of series that we'll be looking at for which we will be able to get actual values for the series.

The third type is divergent and so won't have a value to worry about. In general, determining the value of a series is very difficult and outside of these two kinds of series that we'll look at in this section we will not be determining the value of series in this chapter. So, let's get started.

GEOMETRIC SERIES

A geometric series is any series that can be written in the form,

$$\sum_{n=1}^{\infty} ar^{n-1}$$

or, with an index shift the geometric series will often be written as,

$$\sum_{n=0}^{\infty} ar^{n}$$

These are identical series and will have identical values, provided they converge of course.

If we start with the first form it can be shown that the partial sums are,

$$s_n = \frac{a(1-r^n)}{1-r} = \frac{a}{1-r} - \frac{ar^n}{1-r}$$

The series will converge provided the partial sums form a convergent sequence, so let's take the limit of the partial sums.

$$\lim_{n\to\infty} s_n = \lim_{n\to\infty}\left(\frac{a}{1-r}-\frac{ar^n}{1-r}\right)$$

$$= \lim_{n\to\infty}\frac{a}{1-r} - \lim_{n\to\infty}\frac{ar^n}{1-r}$$

$$= \frac{a}{1-r} - \frac{a}{1-r}\lim_{n\to\infty} r^n$$

Now, from Theorem from the Sequences section we know that the limit above will exist and be finite provided. $-1 < r \le 1$ However, note that we can't let $r = 1$ since this will give division by zero. Therefore, this will exist and be finite provided $-1 < r < 1$ and in this case the limit is zero and so we get,

$$\lim_{n\to\infty} s_n = \frac{a}{1-r}$$

Therefore, a geometric series will converge if $-1 < r < 1$, which is usually written $|r| < 1$, its value is,

$$\sum_{n=1}^{\infty} ar^{n-1} = \sum_{n=0}^{\infty} ar^n = \frac{a}{1-r}$$

Note that in using this formula we'll need to make sure that we are in the correct form. In other words, if the series starts at $n = 0$ then the exponent on the r must be n. Likewise if the series starts at $n = 1$ then the exponent on the r must be $n - 1$.

Example: Determine if the following series converge or diverge. If they converge give the value of the series.

1. $\sum_{n=1}^{\infty} 9^{-n+2}4^{n+1}$

2. $\sum_{n=0}^{\infty} \frac{(-4)^{3n}}{5^{n-1}}$

Solution:

1. $\sum_{n=1}^{\infty} 9^{-n+2}4^{n+1}$

This series doesn't really look like a geometric series. However, notice that both parts of the series term are numbers raised to a power. This means that it can be put into the form of a geometric series. We will just need to decide which form is the correct form. Since the series starts at $n = 1$ we will want the exponents on the numbers to be $n - 1$.

It will be fairly easy to get this into the correct form. Let's first rewrite things slightly. One of the n's in the exponent has a negative in front of it and that can't be there in the geometric form. So, let's first get rid of that.

$$\sum_{n=1}^{\infty} 9^{-n+2} 4^{n+1} = \sum_{n=1}^{\infty} 9^{-(n+2)} 4^{n+1} = \sum_{n=1}^{\infty} \frac{4^{n+1}}{9^{n-2}}$$

Now let's get the correct exponent on each of the numbers. This can be done using simple exponent properties.

$$\sum_{n=1}^{\infty} 9^{-n+2} 4^{n+1} = \sum_{n=1}^{\infty} \frac{4^{n+1}}{9^{n-2}} = \sum_{n=1}^{\infty} \frac{4^{n-1} 4^{2}}{9^{n-1} 9^{-1}}$$

Now, rewrite the term a little.

$$\sum_{n=1}^{\infty} 9^{-n+2} 4^{n+1} = \sum_{n=1}^{\infty} 16(9) \frac{4^{n-1}}{9^{n-1}} = \sum_{n=1}^{\infty} 144 \left(\frac{4}{9}\right)^{n-1}$$

So, this is a geometric series with $a = 144$ and $r = \frac{4}{9} < 1$. Therefore, since $|r| < 1$ we know the series will converge and its value will be,

$$\sum_{n=1}^{\infty} 9^{-n+2} 4^{n+1} = \frac{144}{1 - \frac{4}{9}} = \frac{9}{5}(144) = \frac{1296}{5}$$

$$\sum_{n=1}^{\infty} 9^{-n+2} 4^{n+1} = \frac{144}{1 - \frac{4}{9}} = \frac{9}{5}(144) = \frac{1296}{5}$$

1. $\displaystyle\sum_{n=0}^{\infty} \frac{(-4)^{3n}}{5^{n-1}}$

Again, this doesn't look like a geometric series, but it can be put into the correct form. In this case the series starts at $n = 0$ so we'll need the exponents to be n on the terms. Note that this means we're going to need to rewrite the exponent on the numerator a little,

$$\sum_{n=0}^{\infty} \frac{(-4)^{3n}}{5^{n-1}} = \sum_{n=0}^{\infty} \frac{\left((-4)^3\right)^n}{5^n 5^{-1}} = \sum_{n=0}^{\infty} 5\frac{(-64)^n}{5^n} = \sum_{n=0}^{\infty} 5\left(\frac{-64}{5}\right)^n$$

So, we've got it into the correct form and we can see that $a =$ 5 and $r = -\frac{64}{5}$. Also note that $|r| \geq 1$ and so this series diverges.

Example: Use the results from the previous example to determine the value of the following series.

1. $\sum_{n=0}^{\infty} 9^{-n+2} 4^{n+1}$

2. $\sum_{n=3}^{\infty} 9^{-n+2} 4^{n+1}$

Solution:

1. $\sum_{n=0}^{\infty} 9^{-n+2} 4^{n+1}$

In this case we could just acknowledge that this is a geometric series that starts at $n = 0$ and so we could put it into the correct form and be done with it. However, this does provide us with a nice example of how to use the idea of stripping out terms to our advantage. Let's notice that if we strip out the first term from this series we arrive at,

$$\sum_{n=0}^{\infty} 9^{-n+2} 4^{n+1} = 9^2 4^1 + \sum_{n=1}^{\infty} 9^{-n+2} 4^{n+1} = 324 + \sum_{n=1}^{\infty} 9^{-n+2} 4^{n+1}$$

From the previous example we know the value of the new series that arises here and so the value of the series in this example is,

$$\sum_{n=1}^{\infty} 9^{-n+2} 4^{n+1} = 324 + \frac{1296}{5} = \frac{2916}{5}$$

2. $\sum_{n=3}^{\infty} 9^{-n+2} 4^{n+1}$

In this case we can't strip out terms from the given series to arrive at the series used in the previous example. However, we can start with the series used in the previous example and strip terms out of it to get the series in this example. So, let's do that. We will strip out the first two terms from the series we looked at in the previous example.

$$\sum_{n=1}^{\infty} 9^{-n+2} 4^{n+1} = 9^1 4^2 + 9^0 4^3 + \sum_{n=3}^{\infty} 9^{-n+2} 4^{n+1}$$

$$= 208 + \sum_{n=3}^{\infty} 9^{-n+2} 4^{n+1}$$

We can now use the value of the series from the previous example to get the value of this series.

$$\sum_{n=3}^{\infty} 9^{-n+2} 4^{n+1} = \sum_{n=1}^{\infty} 9^{-n+2} 4^{n+1} - 208 = \frac{1296}{5} - 208 = \frac{266}{5}$$

Notice that we didn't discuss the convergence of either of the series in the above example. Here's why. Consider the following series written in two separate ways (*i.e.* we stripped out a couple of terms from it).

$$\sum_{n=0}^{\infty} a_n = a_0 + a_1 + a_2 + \sum_{n=3}^{\infty} a_n$$

Let's suppose that we know $\sum_{n=3}^{\infty} a_n$ is a convergent series. This means that it's got a finite value and adding three finite terms onto this will not change that fact. So the value of $\sum_{n=0}^{\infty} a_n$ is also finite and so is convergent.

Likewise, suppose that $\sum_{n=0}^{\infty} a_n$ is convergent. In this case if we subtract three finite values from this value we will remain finite and arrive at the value of $\sum_{n=3}^{\infty} a_n$. This is now a finite value and so this series will also be convergent. In other words, if we have two series and they differ only by the presence, or absence, of a finite number of finite terms they will either both be convergent or they will both be divergent.

The difference of a few terms one way or the other will not change the convergence of a series. This is an important idea and we will use it several times in the following sections to simplify some of the tests that we'll be looking at.

TELESCOPING SERIES

It's now time to look at the second of the three series in this section. In this portion we are going to look at a series that is called a telescoping series. The name in this case comes from what happens with the partial sums and is best shown in an example.

Example: Determine if the following series converges or diverges. If it converges find its value.

$$\sum_{n=0}^{\infty} \frac{1}{n^2+3n+2}$$

Solution: We first need the partial sums for this series.

$$sn = \sum_{n=0}^{\infty} \frac{1}{i^2+3i+2}$$

Now, let's notice that we can use partial fractions on the series term to get,

$$\frac{1}{i^2+3i+2} = \frac{1}{(i+2)(i+1)} = \frac{1}{i+1} - \frac{1}{i+2}$$

I'll leave the details of the partial fractions to you. By now you should be fairly adept at this since we spent a fair amount of time doing partial fractions back in the Integration Techniques chapter. If you need a refresher you should go back and review that section. So, what does this do for us? Well, let's start writing out the terms of the general partial sum for this series using the partial fraction form.

$$s_n = \sum_{i=0}^{n}\left(\frac{1}{i+1}-\frac{1}{i+2}\right)$$

$$=\left(\frac{1}{1}-\frac{1}{2}\right)+\left(\frac{1}{2}-\frac{1}{3}\right)+\left(\frac{1}{3}-\frac{1}{4}\right)$$

$$+\ldots+\left(\frac{1}{n}-\frac{1}{n+1}\right)+\left(\frac{1}{n+1}-\frac{1}{n+2}\right)$$

$$=1-\frac{1}{n+2}$$

Notice that every term except the first and last term canceled out. This is the origin of the name telescoping series. This also means that we can determine the convergence of this series by taking the limit of the partial sums.

$$\lim_{n\to\infty} s_n = \lim_{n\to\infty}\left(1-\frac{1}{n+2}\right)=1$$

The sequence of partial sums is convergent and so the series is convergent and has a value of

$$\sum_{n=0}^{\infty}\frac{1}{n^2+3n+2}=1$$

In telescoping series be careful to not assume that successive terms will be the ones that cancel. Consider the following example.

Example: Determine if the following series converges or diverges. If it converges find its value.

$$\sum_{n=1}^{\infty}\frac{1}{n^2+4n+3}$$

Solution: As with the last example we'll leave the partial fractions details to you to verify.

The partial sums are,

$$s_n = \sum_{n=1}^{\infty}\left(\frac{\frac{1}{2}}{i+1} - \frac{\frac{1}{2}}{i+3}\right) = \frac{1}{2}\sum_{n=1}^{\infty}\left(\frac{\frac{1}{2}}{i+1} - \frac{\frac{1}{2}}{i+3}\right)$$

$$= \frac{1}{2}\left[\left(\frac{1}{2}-\frac{1}{4}\right)+\left(\frac{1}{3}-\frac{1}{5}\right)+\left(\frac{1}{4}-\frac{1}{6}\right)\right.$$

$$\left.+\ldots+\left(\frac{1}{n}-\frac{1}{n+2}\right)+\left(\frac{1}{n+1}-\frac{1}{n+3}\right)\right]$$

$$= \frac{1}{2}\left[\frac{1}{2}+\frac{1}{3}-\frac{1}{n+2}-\frac{1}{n+3}\right]$$

In this case instead of successive terms canceling a term will cancel with a term that is farther down the list. The end result this time is two initial and two final terms are left.

Notice as well that in order to help with the work a little we factored the out of the series.

The limit of the partial sums is,

$$\lim_{n\to\infty} s_n$$

$$= \lim_{n\to\infty}\frac{1}{2}\left(\frac{5}{6}-\frac{1}{n+2}-\frac{1}{n+3}\right)=\frac{5}{12}$$

So, this series is convergent (because the partial sums form a convergent sequence) and its value is,

$$\sum_{n=1}^{\infty}\frac{1}{n^2+4n+3}=\frac{5}{12}$$

Note that it's not always obvious if a series is telescoping or not until you try to get the partial sums and then see if they are in fact telescoping. There is no test that will tell us that we've got a telescoping series right off the bat.

Also note that just because you can do partial fractions on a series term does not mean that the series will be a telescoping series.

The following series, for example, is not a telescoping series despite the fact that we can partial fraction the series terms.

$$\sum_{n=1}^{\infty} \frac{3+2n}{n^2+3n+2} = \sum_{n=1}^{\infty}\left(\frac{1}{n+1}+\frac{1}{n+2}\right)$$

In order for a series to be a telescoping we must get terms to cancel and all of these terms are positive and so none will cancel.

Next, we need to go back and address an issue that was first raised in the previous section. In that section we stated that the sum or difference of convergent series was also convergent and that the presence of a multiplicative constant would not affect the convergence of a series.

Now that we have a few more series in hand let's work a quick example showing that.

Example: Determine the value of the following series.

$$\sum_{n=1}^{\infty}\left(\frac{4}{n^2+4n+3} - 9^{-n+2}4^{n+1}\right)$$

Solution: To get the value of this series all we need to do is rewrite it and then use the previous results.

$$\sum_{n=1}^{\infty}\left(\frac{4}{n^2+4n+3} - 9^{-n+2}4^{n+1}\right) = \sum_{n=1}^{\infty}\frac{4}{n^2+4n+3} - \sum_{n=1}^{\infty}9^{-n+2}4^{n+1}$$

$$= 4\sum_{n=1}^{\infty}\frac{4}{n^2+4n+3} - \sum_{n=1}^{\infty}9^{-n+2}4^{n+1}d$$

$$= 4\left(\frac{5}{12}\right) - \frac{1296}{5}$$

$$= \frac{3863}{15}$$

We didn't discuss the convergence of this series because it was the sum of two convergent series and that guaranteed that the original series would also be convergent.

HARMONIC SERIES

This is the third and final series that we're going to look at in this chapter. Here is the harmonic series.

$$\sum_{n=1}^{\infty} \frac{1}{n}$$

The harmonic series is divergent and we'll need to wait until the next section to show that. This series is here because it's got a name and so I wanted to put it here with the other two named series that we looked at in this section. We're also going to use the harmonic series to illustrate a couple of ideas about divergent series that we've already discussed for convergent series. We'll do that with the following example.

Example: Show that each of the following series are divergent.

1. $\sum_{n=1}^{\infty} \frac{5}{n}$

2. $\sum_{n=1}^{\infty} \frac{1}{n}$

Solution:

1. $\sum_{n=1}^{\infty} \frac{5}{n}$

To see that this series is divergent all we need to do is use the fact that we can factor a constant out of a series as follows,

$$\sum_{n=1}^{\infty} \frac{5}{n} = 5 \sum_{n=1}^{\infty} \frac{1}{n}$$

Now, $\sum_{n=1}^{\infty} \frac{1}{n}$ is divergent and so five times this will still not be a finite number and so the series has to be divergent. In other words, if we multiply a divergent series by a constant it will still be divergent.

2. $\sum_{n=4}^{\infty} \frac{1}{n}$

In this case we'll start with the harmonic series and strip out the first three terms.

$$\sum_{n=1}^{\infty}\frac{1}{n}=1+\frac{1}{2}+\frac{1}{3}+\sum_{n=1}^{\infty}\frac{1}{n} \qquad \Rightarrow \qquad \sum_{n=4}^{\infty}\frac{1}{n}=\left(\sum_{n=4}^{\infty}\frac{1}{n}\right)-\frac{11}{6}$$

In this case we are subtracting a finite number from a divergent series. This subtraction will not change the divergence of the series. We will either have infinity minus a finite number, which is still infinity, or a series with no value minus a finite number, which will still have no value. Therefore, this series is divergent. Just like with convergent series, adding/subtracting a finite number from a divergent series is not going to change the fact the convergence of the series.

INTEGRAL TEST

The last topic that we discussed in the previous section was the harmonic series. In that discussion we stated that the harmonic series was a divergent series. It is now time to prove that statement. This proof will also get us started on the way to our next test for convergence that we'll be looking at.

So, we will be trying to prove that the harmonic series,

$$\sum_{n=1}^{\infty}\frac{1}{n}$$

diverges. We'll start this off by looking at an apparently unrelated problem. Let's start off by asking what the area under $f(x)=\frac{1}{x}$ is on the interval $[1,\infty)$. From the section on Improper Integrals we know that this is,

$$\int_{1}^{\infty}\frac{1}{x}\,dx=\infty$$

and so we called this integral divergent (yes, that's the same term we're using here with series). So, just how does that help us to prove that the harmonic series diverges? Well, recall that we can always estimate the area by breaking up the interval into segments and then sketching in rectangles and using the sum of the area all of the rectangles as an estimate of the actual area. Let's do that for this problem as well and see what we get. We will break up the interval into subintervals of width 1 and we'll take the function value at the left endpoint as the height of the rectangle. The image below shows the first few rectangles for this area.

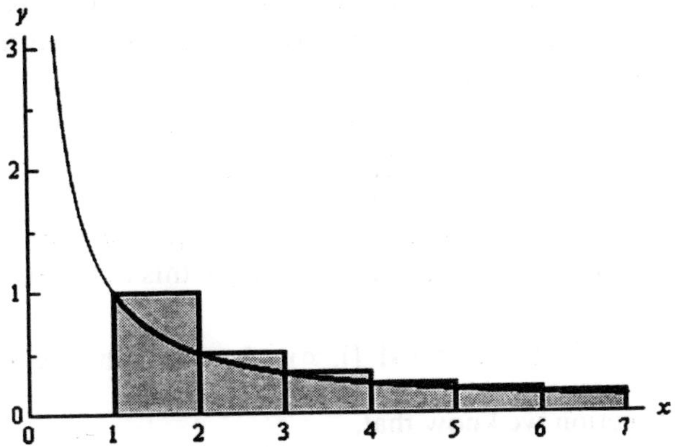

So, the area under the curve is approximately,

$$\begin{aligned} A &\approx \left(\frac{1}{1}\right)(1)+\left(\frac{1}{2}\right)(1)+\left(\frac{1}{3}\right)(1)+\left(\frac{1}{4}\right)(1)+\left(\frac{1}{5}\right)(1)+\ldots \\ &= \frac{1}{1}+\frac{1}{2}+\frac{1}{3}+\frac{1}{4}+\frac{1}{5}+\ldots \\ &= \sum_{n=1}^{\infty}\frac{1}{n} \end{aligned}$$

Now note a couple of things about this approximation. First, each of the rectangles overestimates the actual area and secondly the formula for the area is exactly the harmonic series!

Putting these two facts together gives the following,

$$A \approx \sum_{n=1}^{\infty}\frac{1}{n} > \int_{1}^{\infty}\frac{1}{x}\,dx = \infty$$

Notice that this tells us that we must have,

$$\sum_{n=1}^{\infty}\frac{1}{n} > \infty \quad \Rightarrow \quad \sum_{n=1}^{\infty}\frac{1}{n}\;\infty$$

Since we can't really be larger than infinity the harmonic series must also be infinite in value. In other words, the harmonic series is in fact divergent. So, we've managed to relate a series to an improper integral that we could compute and it turns out that the improper integral and the series have exactly the same convergence. Let's see if this will also be true for a series that converges. When discussing the Divergence Test we made the claim that

$$\sum_{n=1}^{\infty} \frac{1}{n^2}$$

converges. Let's see if we can do something similar to the above process to prove this. We will try to relate this to the area under $f(x) = \frac{1}{x^2}$ is on the interval $[1, \infty)$. Again, from the Improper Integral section we know that,

$$\int_1^{\infty} \frac{1}{x^2}\, dx = 1$$

and so this integral converges. We will once again try to estimate the area under this curve. We will do this in an almost identical manner as the previous part with the exception that we will instead of using the left end points for the height of our rectangles we will use the right end points. Here is a sketch of this case,

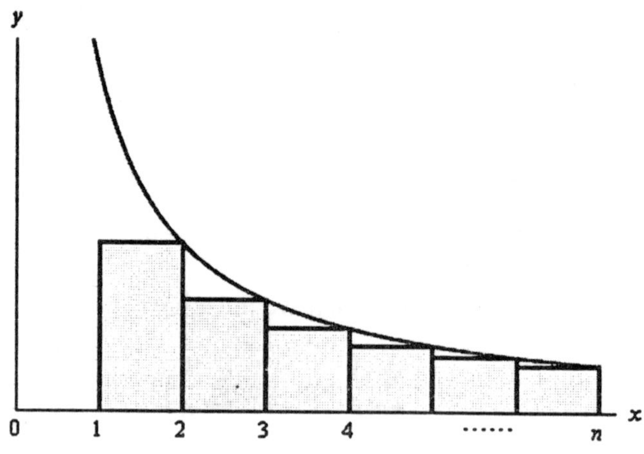

In this case the area estimation is,

$$A \approx \left(\frac{1}{2^2}\right)(1)+\left(\frac{1}{3^2}\right)(1)+\left(\frac{1}{4^2}\right)(1)+\left(\frac{1}{5^2}\right)(1)+\ldots$$
$$=\frac{1}{2^2}+\frac{1}{3^2}+\frac{1}{4^2}+\frac{1}{5^2}+\ldots$$

This time, unlike the first case, the area will be an underestimation of the actual area and the estimation is not quite the series that we are working with. Notice however that the only difference is that we're missing the first term. This means we can do the following,

$$\sum_{n=4}^{\infty}\frac{1}{n^2}=\sum_{n=4}^{\infty}\frac{1}{n^2}\frac{1}{1^2}+\underbrace{\frac{1}{2^2}+\frac{1}{3^2}+\frac{1}{4^2}+\frac{1}{5^2}+\ldots}_{\text{Area Estimation}}<1+\int_1^{\infty}\frac{1}{x^2}\,dx$$
$$=1+1=2$$

Or, putting all this together we see that,

$$\sum_{n=4}^{\infty}\frac{1}{n^2}<2$$

With the harmonic series this was all that we needed to say that the series was divergent. With this series however, this isn't quite enough. For instance $-\infty<2$ and if the series did have a value of $-\infty$ then it would be divergent (when we want convergent).So, let's do a little more work.

First, let's notice that all the series terms are positive (that's important) and that the partial sums are,

$$s_n=\sum_{n=1}^{n}\frac{1}{i^2}$$

Because the terms are all positive we know that the partial sums must be an increasing sequence. In other words,

$$s_n=\sum_{n=1}^{n}\frac{1}{i^2}<\sum_{n=1}^{n+1}\frac{1}{i^2}=s_{n+1}$$

In s_{n+1} we are adding a single positive term onto s_n and so must get larger. Therefore, the partial sums form an increasing (and hence monotonic) sequence. Also note that, since the terms are all positive, we can say,

$$s_n = \sum_{i=1}^{n} \frac{1}{i^2} = < \sum_{i=1}^{\infty} \frac{1}{n^2} < 2 \qquad \Rightarrow \qquad s_n < 2$$

and so the sequence of partial sums is a bounded sequence.

In the second section on Sequences we gave a theorem that stated that a bounded and monotonic sequence was guaranteed to be convergent. This means that the sequence of partial sums is a convergent sequence. So, who cares right? Well recall that this means that the series must then also be convergent! So, once again we were able to relate a series to an improper integral (that we could compute) and the series and the integral had the same convergence.

We went through a fair amount of work in both of these examples to determine the convergence of the two series. Luckily for us we don't need to do all this work every time. The ideas in these two examples can be summarized in the following test.

INTEGRAL TEST

Suppose that $f(x)$ is a continuous, positive and decreasing function on the interval $[k, \infty)$ and that $f(n) = a_n$ then,

1. If $\int_k^{\infty} f(x)dx$ is convergent so is $\sum_{n=k}^{\infty} a_n$.

2. If $\int_k^{\infty} f(x)dx$ is divergent so is $\sum_{n=k}^{\infty} a_n$.

A formal proof of this test can be found at the end of this section. There are a couple of things to note about the integral test. First, the lower limit on the improper integral must be the same value that starts the series.

Second, the function does not actually need to be decreasing and positive everywhere in the interval. All that's really required is that eventually the function is decreasing and positive. In other words, it is okay if the function (and hence series terms) increases or is negative for a while, but eventually the function (series terms) must decrease and be positive for all terms. To see why this is true let's suppose that the series terms increase and or are negative in the range $k \le n \le N$ and then decrease and are positive for $n \ge N + 1$. In this case the series can be written as,

$$\sum_{n=k}^{\infty} a_n = \sum_{n=k}^{N} a_n + \sum_{n=N+1}^{\infty} a_n$$

Now, the first series is nothing more than a finite sum (no matter how large N is) of finite terms and so will be finite. So the original series will be convergent/divergent only if the second infinite series on the right is convergent/divergent and the test can be done on the second series as it satisfies the conditions of the test. A similar argument can be made using the improper integral as well. The requirement in the test that the function/series be decreasing and positive everywhere in the range is required for the proof.

In practice however, we only need to make sure that the function/series is eventually a decreasing and positive function/ series. Also note that when computing the integral in the test we don't actually need to strip out the increasing/negative portion since the presence of a small range on which the function is increasing/ negative will not change the integral from convergent to divergent or from divergent to convergent.

There is one more very important point that must be made about this test. This test does NOT give the value of a series. It will only give the convergence/divergence of the series. That's it. No value. We can use the above series as a perfect example of this. All that the test gave us was that,

$$\sum_{n=1}^{\infty} \frac{1}{n^2} < 2$$

So, we got an upper bound on the value of the series, but not an actual value for the series. In fact, from this point on we will not be asking for the value of a series we will only be asking whether a series converges or diverges. In a later section we look at estimating values of series, but even in that section still won't actually be getting values of series. Just for the sake of completeness the value of this series is known.

$$\sum_{n=1}^{\infty} \frac{1}{n^2} = \frac{\pi^2}{6} = 1.644934... < 2$$

Let's work a couple of examples.

Example: Determine if the following series is convergent or divergent.

$$\sum_{n=2}^{\infty} \frac{1}{n \text{ In } n}$$

Solution: In this case the function we'll use is,

$$f(x) = \frac{1}{x \text{ In } x}$$

This function is clearly positive and if we make x larger the denominator will get larger and so the function is also decreasing. Therefore, all we need to do is determine the convergence of the following integral.

$$\int_2^{\infty} \frac{1}{x \text{ In } x} dx = \lim_{n \to \infty} \int_2^{\infty} \frac{1}{x \text{ In } x} dx \qquad u = \text{In } x$$

$$= \lim_{n \to \infty} \left(\text{In } (\text{In } x) \right)\Big|_2^t$$

$$= \lim_{n \to \infty} \left(\text{In } (\text{In } t) - \text{In } (\text{In } 2) \right)$$

$$= \infty$$

The integral is divergent and so the series is also divergent by the Integral Test.

Example: Determine if the following series is convergent or divergent.

$$\sum_{n=0}^{\infty} ne^{-n^2}$$

Solution: The function that we'll use in this example is,

$$f(x) = xe^{-x^2}$$

This function is always positive on the interval that we're looking at. Now we need to check that the function is decreasing. It is not clear that this function will always be decreasing on the interval given.

We can use our Calculus I knowledge to help us however. The derivative of this function is,

$$f'(x) = e^{-x^2}(1-2x^2)$$

This function has two critical points (which will tell us where the derivative changes sign) at $x = \pm\frac{1}{\sqrt{2}}$.

Since we are starting at $n = 0$ we can ignore the negative critical point. Picking a couple of test points we can see that the function in increasing on the interval $\left[0, \frac{1}{\sqrt{2}}\right]$ and it is decreasing on $\left[\frac{1}{\sqrt{2}}, \infty\right)$.

Therefore, eventually the function will be decreasing and that's all that's required for us to use the Integral Test.

$$\int_0^\infty xe^{-x^2}\,dx \quad =$$

$$\lim_{x\to\infty}\int_0^\infty xe^{-x^2}\,dx \qquad u = -x^2$$

$$= \lim_{x\to\infty}\left(-\frac{1}{2}xe^{-x^2}\right)\Bigg|_0^t$$

$$= \lim_{x\to\infty}\left(\frac{1}{2}-\frac{1}{2}e^{-t^2}\right) = \frac{1}{2}$$

The integral is convergent and so the series must also be convergent by the Integral Test.

FACT (THE *P* – SERIES TEST)

If $k > 0$ then $\sum_{n=k}^{\infty}\frac{1}{n^p}$ converges if $p > 1$ and diverges if $p \leq 1$.

Sometimes the series in this fact are called p-series and so this fact is sometimes called the p-series test. This fact follows directly from the Integral Test and a similar fact we say in the Improper Integral section. This fact says that the integral,

$$\int_k^{\infty} \frac{1}{x^p}\,dx$$

converges if $p > 1$ and diverges if $p \leq 1$.

Using the *p*-series test makes it very easy to determine the convergence of some series.

Example: Determine if the following series are convergent or divergent.

1. $\sum_{n=4}^{\infty} \frac{1}{n^7}$

2. $\sum_{n=1}^{\infty} \frac{1}{\sqrt{n}}$

Solution:

1. In this case $p = 7 > 1$ and so by this fact the series is convergent.
2. For this series $p = \frac{1}{2} \leq 1$ and so the series is divergent by the fact.

The last thing that we'll do in this section is give a quick proof of the Integral Test.

We've essentially done the proof already at the beginning of the section when we were motivating the Integral Test, but let's go through it formally for a general function.

Proof of Integral Test

First, for the sake of the proof we'll be working with the series $\sum_{n=4}^{\infty} a_n$. The original test statement was for a series that started at a general $n = k$ and while the proof can be done for that it will be easier if we assume that the series starts at $n = 1$. Another way of dealing with the $n = k$ is we could do an index shift and start the series at $n = 1$ and then do the Integral Test. Either way proving the test for $n = 1$ will be sufficient. Let's start off and estimate the area under the curve on the interval $[1, n]$ and we'll underestimate the area by taking rectangles of width one and whose height is the right endpoint. This gives the following figure.

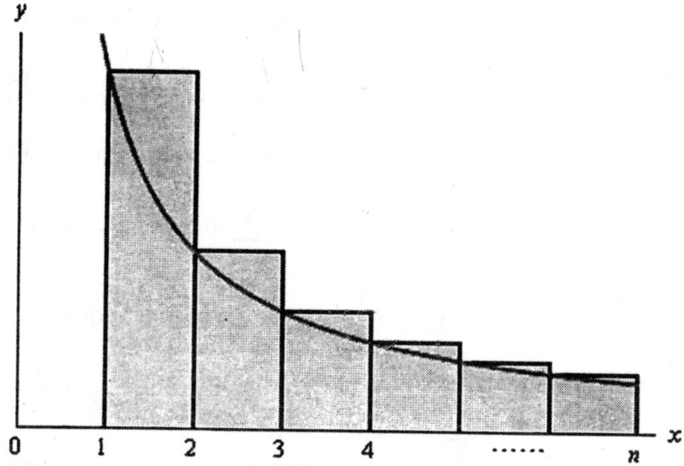

Now, note that,

$f(2) = a_2 \qquad f(3) = a_3 \qquad \dots \qquad f(n) = a_n$

The approximate area is then,

$A \approx (1)f(2) + (1)f(3) + \dots + (1)f(n) = a_2 + a_3 + \dots a_n$

and we know that this underestimates the actual area so,

$$\sum_{n=4}^{\infty} a_i = a_2 + a_3 + \dots a_n < \int_1^n f(x)\,dx$$

Now, let's suppose that $\int_1^\infty f(x)\,dx$ is convergent and so $\int_1^\infty f(x)\,dx$ must have a finite value. Also, because $f(x)$ is positive we know that, $\int_1^\infty f(x)\,dx < \int_1^\infty f(x)\,dx$

This in turn means that,

$$\sum_{i=2}^{n} a_i < \int_1^n f(x)\,dx < \int_1^\infty f(x)\,dx$$

Our series starts at $n = 1$ so this isn't quite what we need. However, that's easy enough to deal with.

$$\sum_{i=2}^{n} a_i = a_1 + \sum_{i=2}^{n} a_i < a_1 + \int_1^{\infty} f(x)\,dx = M$$

So, just what has this told us? Well we now know that the sequence of partial sums, $sn = \sum_{i=4}^{n} a_i$ are bounded above by M.

Next, because the terms are positive we also know that

$$s_n \le s_n + a_{n+1} = \sum_{i=4}^{n} a_i + a_{n+1} = \sum_{i=4}^{n+1} a_i + s_{n+1} \Rightarrow s_n \le s_{n+4}$$

and so the sequence $\{s_n\}_{n=1}^{\infty}$ is also an increasing sequence. So, we now know that the sequence of partial sums $\{s_n\}_{n=1}^{\infty}$ converges and hence our series $\sum_{i=4}^{\infty} a_n$ is convergent.

So, the first part of the test is proven. The second part is somewhat easier. This time let's over estimate the area under the curve by using the left endpoints of interval for the height of the rectangles as shown below.

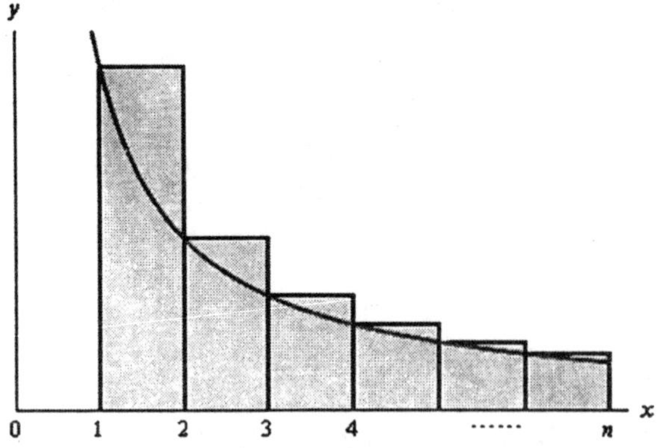

In this case the area is approximately,

$$A \gg (1)f(1) + (1)f(2) + ... + (1)f(n-1) = a_1 + a_2 + ...\ a_{n-1}$$

Since we know this overestimates the area we also then know that,

$$s_{n-1}\sum_{i=1}^{n-1} a_i = a_1 + a_2 + ...a_{n-1} > \int_1^{n-1} f(x)\,dx$$

Now, suppose that $\int_1^{\infty} f(x)\,dx$ is divergent. In this case this means that $\int_1^{\infty} f(x)\,dx \to \infty$ as because $f(x) \geq 0$.

However, because $n - 1 \to \infty$ as $n \to \infty$ we also know that $\int_1^{n-1} f(x)\,dx \to \infty$.

Therefore, since $s_{n-1} > \int_1^{n-1} f(x)\,dx$ we know that as $n \to \infty$ we must have $s_{n-1} \to \infty$. This in turn tells us that $s_n \to \infty$ as $n \to \infty$.

So, we now know that the sequence of partial sums, $\{s_n\}_{n=1}^{\infty}$, is a divergent sequence and so $\sum_{n=1}^{\infty} a_n$ is a divergent series.

1. Another example is given by the function $f(x) = e^{-1/x2}$ for x ψ0, which, in spite of its apparently regular behaviour, cannot be expanded in a Taylor series. As a matter of fact, this function ceases to be continuous, if we take pure imaginary values of $x = i\xi$ into account. The function then takes the form e^{1/ξ^2} and increases beyond all bounds as ξ } 0.

It is therefore clear that no power series in x can represent this function for all complex values of x in a neighbourhood of the origin, no matter how small a neighbourhood we choose. These remarks on the theory of functions and power series of a complex variable must be sufficient for us here.

3

Infinite Series

The idea of an infinite series expansion of a function was first conceived in India by Madhava in the 14th century, who also developed the concepts of the power series, the Taylor series, the Maclaurin series, rational approximations of infinite series, and infinite continued fractions.

He discovered a number of infinite series, including the Taylor series of the trigonometric functions of sine, cosine, tangent and arctangent, the Taylor series approximations of the sine and cosine functions, and the power series of the radius, diameter, circumference, angle θ, π and $\pi/4$. His students and followers in the Kerala School further expanded his works with various other series expansions and approximations, until the 16th century. In the 17th century, James Gregory also worked on infinite series and published several Maclaurin series. In 1715, a general method for constructing the Taylor series for all functions for which they exist was provided by Brook Taylor.

Leonhard Euler in the 18th century, developed the theory of hypergeometric series and q-series. In Mathematics a Series is often represented as the sum of a sequence of terms. That is, a series is represented as a list of numbers with addition operations between them, for example this arithmetic sequence:

$$1 + 2 + 3 + 4 + 5 + \ldots + 99 + 100$$

In most cases of interest the terms of the sequence are produced according to a certain rule, such as by a formula, by an algorithm, by a sequence of measurements, or even by a random number generator. A series may be finite or *infinite*. Finite series may be handled with elementary algebra, but infinite series require tools from mathematical analysis if they are to be applied in anything more than a tentative way. Examples of simple series include the arithmetic series which is a sum of an arithmetic progression, written as:

$$\sum_{n=0}^{k}(an+b);$$

and finite geometric series, a sum of a geometric progression, which can be written as:

$$\sum_{n=0}^{k}a^n.$$

INFINITE SERIES

The sum of an infinite series $a_0 + a_1 + a_2 + \ldots$ is the limit of the sequence of partial sums

$$S_n = a_0 + a_1 + a_2 + \ldots + a_n$$

as $n \to \infty$, if that limit exists. If the limit exists and is finite, the series is said to converge; if it is infinite or does not exist, the series is said to diverge.

The easiest way that an infinite series can converge is if all the a_n are zero for n sufficiently large. Such a series can be identified with a finite sum, so it is only infinite in a trivial sense. However, infinite series of nonzero terms can also converge, which resolves the mathematical side of several of Zeno's paradoxes. The simplest case of a nontrivial infinite series is perhaps

$$1+\frac{1}{2}+\frac{1}{4}+\frac{1}{8}+\frac{1}{16}+\ldots$$

It is possible to "visualize" its convergence on the real number line: we can imagine a line of length 2, with successive segments marked off of lengths 1, ½, ¼, etc. There is always room to mark the next segment, because the amount of line remaining is always the same as the last segment marked: when we have marked off ½, we still have a piece of length ½ unmarked, so we can certainly mark the next ¼. This argument does not prove that the sum is *equal* to 2 (although it is), but it does prove that it is *at most* 2. In other words, the series has an upper bound. This series is a geometric series and mathematicians usually write it as:

$$\sum_{n=0}^{\infty} 2^{-n} = 2.$$

An infinite series is formally written as

$$\sum_{n=0}^{\infty} a_n$$

where the elements a_n are real (or complex) numbers. We say that this series converges to S, or that its sum is S, if the limit

$$\lim_{N\to\infty} \sum_{n=0}^{N} a_n$$

exists and is equal to S. If there is no such number, then the series is said to *diverge*.

FORMAL DEFINITION

Mathematicians usually study a series as a *pair* of sequences: the sequence of terms of the series: $a_0, a_1, a_2, \ldots$ and the sequence of partial sums $S_0, S_1, S_2, \ldots$, where $S_n = a_0 + a_1 + \ldots + a_n$. The notation

$$\sum_{n=0}^{\infty} a_n$$

represents the above sequence of partial sums, which is always well defined, but which may or may not converge. In the case of convergence (that is, when the sequence of partial sums S_N has a limit), the notation is also used to denote the *limit* of this sequence. To make a distinction between these two completely different objects (sequence vs. numerical value), one may sometimes omit the limits (atop and below the sum's symbol) in the former case, although it is usually clear from the context which one is meant. Also, different notions of convergence of such a sequence do exist (absolute convergence, summability, etc).

In case the elements of the sequence (and thus of the series) are not simple numbers, but, for example, functions, still more types of convergence can be considered.

Mathematicians extend this idiom to other, equivalent notions of series. For instance, when we talk about a recurring decimal, we are talking, in fact, just about the series for which it stands (0.1 + 0.01 + 0.001 + ...).

But because these series always converge to real numbers (because of what is called the completeness property of the real numbers), to talk about the series in this way is the same as to talk about the numbers for which they stand. In particular, it should offend no sensibilities if we make no distinction between 0.111... and $^1/_9$. Less clear is the argument that $9 \times 0.111... = 0.999... = 1$, but it is not untenable when we consider that we can formalize the proof knowing only that limit laws preserve the arithmetic operations.

The geometric series, Taylor series and several special examples, which we have already met, suggest that we may well study from a rather more general point of view those limiting processes under the heading of summation of infinite series. By its nature, any limiting value $S = \lim_{n\to\infty} s_n$ can be written as an infinite series, because, if n takes the values 1, 2, 3,..., we need only set $a_n = s_n - s_{n-1}$ for $n > 1$ and $a_1 = s_1$, in order to obtain $s_n = a_1 + a_2 + ... + a_n$, and the value S thus appears as the limit of s_n - the sum of n terms as n increases. We express this fact by saying that S is the sum of the infinite series $a_1 + a_2 + a_3 + ...$

Thus, an infinite series is simply a way of representing a limit, where each successive approximation is found from the preceding one by addition of one more term. In principle, the expression of a number as a decimal is merely the representation of a number a in the form of an infinite series $a = a_1 + a_2 + a_3 + ...$ where, if $0 < a < 1$, the term a_n is put equal to $a_n\ 10^{-n}$ and a_n is a whole number between 0 and 9, inclusively.

Since every limiting value can be written in the form of an infinite series, it may seem that a special study of series is superfluous. However, in many cases, it happens that limiting values occur naturally in the form of infinite series, which often exhibit particularly simple laws of formation. Of course, it is not true that every series has an easily recognizable law of formation.

For example, the number *p* can certainly be represented as a decimal, yet we know no simple law which enables us to state the value of an arbitrary digit, say its 7000-th decimal. However, if we set aside the representation of π by a decimal and consider instead Gregory's series, we have an expression with a perfectly clear general law of formation.

Infinite series, in which approximations to the limit are found by repeated addition of new terms, are analogous to infinite products, in which the approximations to the limit arise from repeated multiplication by new factors. However, we shall not go deeply into the theory of infinite products.

The Concepts of Convergence and Divergence

The Fundamental Ideas: We consider an infinite series the general term of which we denote by a_n; the series then has the form $a_1 + a_2 + \ldots = \sum_{\nu=1}^{\infty} a_\nu$.

The symbol with the summation sign on the right hand side is merely an abbreviated way of writing the expression on the left hand side.

For formal reasons, we include the possibility that certain of the numbers a_n may be zero. If all a_n vanish onwards from a number N (i.e., when $n > N$), we speak of a terminating series.

If, as n increases, the n-th partial sum $s_n = a_1 + a_2 + \ldots + a_n = \sum_{\nu=1}^{n} a_y$ approaches a limit $S = \lim_{n\to\infty} s_n$

we say that the series converges, and otherwise that it diverges. In the first case, we call S the sum of the series. We have already met with many examples of convergent series; for instance, the $1 + q + q^2 + \ldots$, geometric series which converges to the sum $1/(1-q)$ for $q < 1$, Gregory's series, the series for log 2, the series for e, and others converge. In the language of infinite series, Cauchy's convergence test is expressed as follows:

A necessary and sufficient condition for the convergence of a series is that the number $|s_m - s_n| = |a_{n+1} + a_{n+2} + \ldots + a_m|$

becomes arbitrarily small if m and n are chosen sufficiently large ($m > n$).

In other words: A series converges if and only if the following condition is met: Given a positive number ε, however small, it is possible to choose an index $N=N(\varepsilon)$, which, in general, increases beyond all bounds as $\varepsilon \to 0$, in such a way that the $|s_m - s_n|$ is less than e, provided only that $m > N$ and $n > N$.

We can make the meaning of the convergence test clearer by considering the geometric series with $q = 1/2$. If we choose $\varepsilon = 1/10$, we need only take $N = 4$, because,

$$|s_m - s_n| = \frac{1}{2^n} + \ldots + \frac{1}{2^{m-1}}$$

$$= \frac{1}{2^{n-1}}\left(\frac{1}{2} + \frac{1}{2^2} + \ldots + \frac{1}{2^{m-n}}\right) < \frac{1}{2^{n-1}}$$

and $\frac{1}{2^{n-1}} < \frac{1}{10}, if\ n > 4$.

If we choose ε equal to 1/100, it is sufficient to take 7 as the corresponding value of N, as is easily verified. Obviously, it is a necessary condition for the convergence of a series that $\lim\limits_{n\to\infty} a_n = 0$. Otherwise, undoubtedly, the convergence criterion cannot be fulfilled. However, this necessary condition is by no means sufficient for convergence; on the contrary, it is easy to find infinite series the general term a_n of which approaches 0 as n increases, yet their sum does not exist, because the partial sum s_n increases without limit as n increases.

An example of this is the series $1 + \frac{1}{\sqrt{2}} + \frac{1}{\sqrt{3}} + \ldots + \frac{1}{\sqrt{n}} + \ldots$ with the general term $\frac{1}{\sqrt{n}}$. We immediately see that

$$s_n > \frac{1}{\sqrt{n}} + \ldots + \frac{1}{\sqrt{n}} = \frac{n}{\sqrt{n}} = \sqrt{n}$$

The n-th partial sum increases beyond all bounds as n increases, whence the the series diverges The same is true for the classical example of the harmonic series $1 + \frac{1}{2} + \frac{1}{3} + \frac{1}{4} + \ldots$, for which,

$$a_{n+1} + ... + a_{2n} = \frac{1}{n+1} + ... + \frac{1}{2n} > \frac{1}{2n} + ... + \frac{1}{2n} = \frac{1}{2}$$

Since n and $m = 2n$ can be taken as large as we please, the series diverges, because Cauchy's test is not fulfilled; in fact, the n-th partial sum tends obviously to infinity, since all the terms are positive. On the other hand, the series, formed from the same numbers with alternating signs,

$$1 - \frac{1}{2} + \frac{1}{3} - \frac{1}{4} + \frac{1}{5} - +.. + \frac{(-1)^{n-1}}{n} + ...$$

converges and has the sum log 2.

It is by no means true that in every divergent series s_n tends to +1 or −1. Thus, in the case of the series 1 − 1 + 1 − 1 + 1 + −..., we see that the partial sum s_n has alternately the values 1 and 0, and, on account of this backwards/ forwards oscillation neither approaches a definite limit nor increases numerically beyond all bounds. As the convergence and divergence of an infinite series is concerned, the following fact which, though self-evident, is very important should be noted: The convergence or divergence of a series is not changed by insertion or removal of finite numbers of terms. As far as convergence or divergence is concerned, it does not matter in the least whether we begin the series at the term a_0, or a_1, or a_5, or at any other arbitrarily chosen term.

Absolute Convergence and Conditional Convergence:. The series $1 + \frac{1}{2} + \frac{1}{3} + \frac{1}{4}...$ diverges; but if we change the sign of every second term, the resulting series converges. On the other hand, the geometric series 1-q+$q$ýÿ-$q$ýÿÿýÿÿýÿ converges and has the sum l/(l + q), provided that $0 < q < 1$; on converting all the signs to +, we obtain the series $1 + q + q^2 + q^3 +...$, which also converges and has the sum l/(l - q).

There appears here a distinction which we must examine more closely. With a series all terms of which are positive, there are only two possible cases; either it converges or the partial sum increases beyond all bounds with n. In fact, the partial sums, being a monotonic increasing sequence, must converge, if they remain bounded. Convergence occurs, if the terms approach zero sufficiently rapidly as n increases; on the other hand, divergence occurs, if the terms do not approach zero at all or if they approach zero too slowly. However, in series, where some terms are positive and others negative, the changes of sign may bring about convergence, since a too great increase in the partial sums, due to the positive terms, is compensated by negative terms, as a final result of which a definite limit is approached.

In order to grasp this fact better with a series $\sum_{\nu=1}^{\infty} a_\nu$, which has positive and negative terms, we compare it with the series which has the same terms all with positive signs, that is,

$$|a_1| + |a_2| + \dots = \sum_{\nu=1}^{\infty} |a_\nu|$$

If this series converges, then for sufficiently large values of n and $m > n$, the expression $|a_1| + |a_{n+2}| + \dots + |a_m|$ will certainly be as small as we please; on account of the relation $|a_{n+1} + \dots + a_m| \leq |a_{n+1}| + \dots + |a_m|$, the expression on the left hand side is also arbitrarily small, whence the original series $\sum_{\nu=1}^{\infty} a_\nu$ converges. In this case, the original series is said to be absolutely convergent. Its convergence is due to the numerical smallness of its terms and does not depend on the change of the signs.

On the other hand, if the series with all the terms taken positively diverges and the original series still converges, we say that the original series is conditionally convergent. Conditional convergence results from the terms of opposite signs compensating one another.

For conditional convergence, Leibnitz's convergence test is frequently of use: If the terms of a series have alternating sign and, moreover, their absolute values $|a_n|$ tend monotonically to 0 (so that $|a_{n+1}| < |a_n|$), the series $\sum_{\nu=1}^{\infty} a_\nu$ converges.

In the proof, we assume that $a_1 > 0$, which does not essentially limit the generality of the argument, and write our series in the form $b_1 - b_2 + b_3 - +\ldots$, where all the terms b_n are now positive, b_n tends to 0 and the condition $b_{n+1} < b_n$ is satisfied. If we bracket the terms together in the two ways $b_1 - (b_2 - b_3) - (b_4 + b_5) -\ldots$ and $(b_1 - b_2) + (b_3 + b_4) + (b_5 + b_6) +\ldots$

we see at once that the partial sums satisfy the two relations:

$$s_1 > s_2 > s_3 > \ldots > s_{2m+1} > \ldots$$

$$s_2 < s_4 < s_6 < \ldots < s_{2m} < \ldots$$

On the other hand, $s_{2n} < s_{2n+1} < s_1$ and $s_{2n+1} > s_{2n} > s_2$, whence the odd partial sums $s_1, s_3,\ldots$ form a monotonic decreasing sequence, which in no case drops below the value s_2 and this sequence has a limit L. The even partial sums $s_2, s_4,\ldots$ likewise form a monotonic increasing sequence the terms of which in no case exceed the fixed number s_1, whence this sequence must have a limiting value L'. Since the numbers s_{2n} and s_{2n+1} differ only by the number b_{2n+1}, which approaches 0 as n increases, the limiting values L and L' are equal to each other. In other words, the even and the odd partial sums

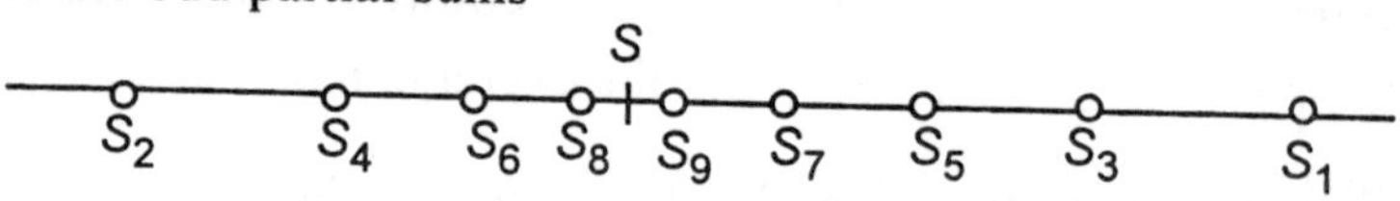

Fig. Convergence of An Alternating Series

approach the same limit, which we will now denote by S. However, this implies that our series is convergent, as has been asserted; its sum is S.

In conclusion, we make another general remark about the fundamental difference between absolute convergence and conditional convergence. We consider a convergent series $\sum_{\nu=1}^{\infty} a_\nu$ and denote the positive terms of the series by $p_1, p_2, p_3,\ldots$ and the negative terms by $-q_1, -q_2, -q_3,\ldots$ If we form the n-th partial sum $s_n = \sum_{\nu=1}^{n} a_\nu$ of the given series, a certain nmnber, say, n' positive terms and a certain number, say n'' negative terms must appear, where $n'+n''=n$. Moreover, if the number of positive terms as well as that of negative terms in the series is infinite, then the two numbers n'' and n'' will increase beyond all bounds with n. We see immediately that the partial sum s_n is simply equal to the partial sum $\sum_{\nu=1}^{n'} p_\nu$ of the positive terms of the series plus the partial sum $-\sum_{\nu=1}^{n''} q_\nu$ of the negative terms.

If the given series converges absolutely, then both the series of positive terms $\sum_{\nu=1}^{\infty} p_\nu$ and the series of the absolute values of the negative terms $\sum_{\nu=1}^{\infty} q_\nu$ certainly converge. In fact, as m increases, the partial sums $\sum_{\nu=1}^{m} p_\nu$ *and* $\sum_{\nu=1}^{m} q_\nu$ are monotonic non-decreasing with the upper bound $\sum_{\nu=1}^{\infty} |a_\nu|$.

The sum of an absolutely convergent series is then simply equal to the sum of the series consisting only of the positive terms plus the sum of the series consisting of the negative terms, or, in other words, it is equal to the difference of the two series with positive terms.

However, $\sum_{\nu=1}^{n} a_\nu = \sum_{\nu=1}^{n'} p_\nu - \sum_{\nu=1}^{n''} q_\nu$; as n increases, n' and n" must also increase beyond all bounds, and the limit of the left hand side must therefore be equal to the difference of the two sums on the right hand side. If the series contains only a finite number of terms of one particular sign, the facts are correspondingly simplified.

On the other hand, if the series does not converge absolutely, but does converge conditionally, then both the series $\sum_{\nu=1}^{\infty} p_\nu$ and

$\sum_{\nu=1}^{\infty} q_{\nu}$ must be divergent, because, if both were convergent, the series would converge absolutely, contrary to our hypothesis. If only one diverges, say $\sum_{\nu=1}^{\infty} p_{\nu}$, and the other converges, then separation into positive and negative parts, $s_n = \sum_{\nu=1}^{n'} p_{\nu} - \sum_{\nu=1}^{n''} q_{\nu}$, shows that the series could not converge; in fact, as n increases, n' and $\sum_{\nu=1}^{n'} p_{\nu}$ would increase beyond all bounds, while the term $\sum_{\nu=1}^{n''} q_{\nu}$ would approach a definite limit, so that the partial sum s_n would increase beyond all bounds.

Hence, we see that a conditionally convergent series cannot be thought of as the difference of two convergent series, one consisting of its positive terms and the other consisting of the absolute values of its negative terms. Closely connected with this fact, there arises another difference between absolutely and conditionally convergent series which we shall now mention briefly.

Rearrangement of Terms

It is a property of finite sums that we may change the order of the terms or, as we say, rearrange the terms at will without changing their values.

There arises the question: What is the exact meaning of a change of the order of terms in an infinite series and does such a rearrangement leave the value of the sum unchanged? While in the case of finite sums there arises no difficulty, for example, in adding the terms in reverse order, in the case of infinite series such a possibility does not exist; there is no last term with which to begin.

Now a change of the order of terms in an infinite series can only mean this: A series $a_1 + a_2 + a_3 +...$ is rearranged into a series $b_1 + b_2 + b_3 + \ldots$, provided that every term a_n of the first series occurs exactly once in the second series and conversely. For example, the amount by which a_n is displaced may increase beyond all bounds as n does; the only point is that it must appear somewhere in the new series.

If some of the terms are moved to later positions in the series, other terms must, of course, be moved to earlier positions. For example, the series $1 + q + q^2 + q^4 + q^3 + q^8 + q^7 + q^6 + q^5 + q^{16} + \ldots$ is a rearrangement of the geometric series $1 + q + q^2 + \ldots$.

As regards the change of order, there is a fundamental distinction between absolutely convergent and conditionally convergent series. In absolutely convergent series, rearrangement of terms does not affect the convergence and the value of the sum of the series is not changed, exactly as in the case of finite sums. On the other hand, in conditionally convergent series, the value of the sum of the series can be changed at will by suitable rearrangement of the series, and the series can even be made to diverge, if so desired. The first of these facts, referring to absolutely convergent series, is easily established. Let us assume, to begin with, that our series has only positive terms and consider the n-th partial sum $s_n = \sum_{\nu=1}^{n} a_\nu$. All the terms of this partial sum occur in the m-th partial sum $t_m = \sum_{\nu=1}^{m} b_\nu$ of the rearranged series, provided only that m is chosen large enough. Hence $t_m \geq s_n$.

On the other hand, we can determine an index n' so large that the partial sum $s_{n'} = \sum_{\nu=1}^{n'} a_\nu$ of the first series contains all the terms b_1, b_2, b_m. It then follows that $t_m \leq s_{n'} \leq A$, where A is the sum of the first series. Thus, for all sufficiently large values of m, we have $s_n \leq t_m \leq A$; since s_n can be made to differ from A by an arbitrarily small amount, it follows that the rearranged series also converges, in fact, to the same limit A as the original series. If the absolutely convergent series has both positive and negative terms, we may regard it as the difference of two series each of which has only positive terms.

Since in the rearrangement of the original series each of these two series merely undergoes rearrangement and therefore converges to the same value as before, the same is true of the original series when rearranged.

In fact, by the case just considered, the new series is absolutely convergent and is therefore the difference of the two rearranged series of positive terms. The fact just proved may seem to the beginner to be a triviality.

The fact that it really does require a proof and that in this proof absolute convergence is essential can be demonstrated by an example of the opposite behaviour of conditionally convergent series.Consider the familiar series

$$1-\frac{1}{2}+\frac{1}{3}-\frac{1}{4}+\frac{1}{5}-\frac{1}{6}+\frac{1}{7}-\frac{1}{8}+-\ldots = \log 2.$$

Write below it the result of multiplication by the factor $1/2\,\frac{1}{2}-\frac{1}{4}+\frac{1}{6}-\frac{1}{8}+-\ldots=\frac{1}{2}\log 2$ and add the two series, combining the terms in vertical columns.

We thus obtain $1-\frac{1}{3}-\frac{1}{2}+\frac{1}{5}+\frac{1}{7}-\frac{1}{4}+\frac{1}{9}+\frac{1}{11}++\ldots=\frac{3}{2}\log 2$.

This last series can obviously be obtained by rearranging the original series and yet the value of the sum of the series has been multiplied by the factor 3/2. It is easy to imagine the effect of the discovery of this apparent paradox on the mathematicians of the Eighteenth Century, who were accustomed to operate with infinite series without regard to their convergence.

We shall give the proof of the theorem stated above concerning the change in the sum of a conditionally convergent series which arises from a change of order of its terms, although we shall have no occasion to make use of the result. Let $p_1, p_2, \ldots$ be the positive terms and $-q_1, -q_2, \ldots$ the negative terms of the series. Since the absolute value $|a_n|$ tends to 0 as n increases, the numbers p_n and q_n must also tend to 0 with n.

Moreover, as we have already seen, the sum $\sum_1^\infty p_\nu$ must diverge and the same is true for $\sum_1^\infty q_\nu$. This abbreviated notation for $\sum_1^\infty p_\nu$ and analogous expressions for other series will be used often in what follows.

We can now find easily a rearrangement of the original series which has an arbitrary number a as its limit. In order to be specific, let a be positive.

We then add the first n_1 positive terms, just enough to ensure that the sum $\sum_1^{n_1} p_\nu$ is less than a. Since the sum $\sum_1^{n_1} p_\nu$ increases with n_1 beyond all bounds, it is always possible, by using enough terms, to make the partial sum larger than a.

The sum will then differ from the exact value a by at most p_n. We now add just enough negative terms $-\sum_1^{m_1} p_\nu$ in order to ensure that the sum $-\sum_1^{m_1} p_\nu$ $-\sum_1^{m_1} q_\nu$ is less than a.

This is also possible, as follows from the divergence of the series from the divergence of the series $\sum_1^\infty q_\nu$.

The difference between this sum and a is now at most q_{m1}. We now add just enough other positive terms $\sum_{n_1+1}^{n_2} p_\nu$ in order to make the partial sum again larger than a, as is again possible since the series of positive terms diverges. The difference between the partial sum and a is now at most p_{ms}.

We again add just just enough negative terms $\sum_{m_1+1}^{m_2} q_\nu$, which start next after the last one previously used, to make the sum once more less than a and continue in the same way.

The values of the sums thus obtained will oscillate about the number a and, when the process is carried out far enough, the oscillation will only take place between arbitrarily narrow bounds; in fact, since the terms p_n and q_n themselves tend to 0 when n is sufficiently large, the length of the interval in which the oscillation occurs will also tend to 0, and the theorem is proved. In the same manner, we can rearrange s series in such a way as to make it diverge; we merely have to choose such large numbers of positive terms as compared with the negative ones that there occurs no longer compensation.

Operations with Infinite Series: It is clear that two convergent infinite series $a_1 + a_2 +... = S$ and $b_1 + b_2 +... = T$ can be added term by term, i.e., that the series formed from the terms $c_n = a_n + b_n$ converges and has the sum $S + T$, because

$$\sum_{\nu=1}^{n} c_\nu = \sum_{\nu=1}^{n} a_\nu + \sum_{\nu=1}^{n} b_\nu \to S + T$$

This theorem is really nothing more than another statement of the fact that the limit of the sum of two terms is the sum of their limits. It is also clear that, if we multiply each term of a convergent infinite series by the same factor, the series remains convergent and its sum is multiplied by the same factor. In the cases just mentioned, it is immaterial whether the convergence is absolute or conditional.

On the other hand, a further study, which is not necessary for us here, shows that, if two infinite series are multiplied by each other by the method used in multiplying finite sums, the product series will not usually converge or have the product of the two sums as its sum unless at least one of the two series is absolutely convergent.

Tests for Convergence and Divergence

We have already encountered a test of a general nature for the convergence of series, which applies to series with terms of alternating signs and decreasing absolute values and which asserts that such series are at least conditionally convergent. In what follows, we shall only consider criteria referring to absolute convergence.

The Comparison Tests

All such considerations of convergence depend on the comparison of the series in question with a second series; this second series is chosen in such a way that its convergence can readily be tested.

The general comparison test may be stated as follows:

If all the numbers b_1, b_2... are positive and the series $\sum_{\nu=1}^{\infty} b_\nu$ converges and if $|a_n| \leqq b_n$ for all values of n, then the series $\sum_{\nu=1}^{\infty} a_n$ is absolutely convergent. If we apply Cauchy's test, the proof becomes almost trivial, because, if m n, we have

$$|a_n + ... + a_m| \leqq |a_n| + ... + |a_m| \leqq b_n + ... + b_m .$$

Since the series $\sum_{\nu=1}^{\infty} b_\nu$ converges, the right hand side is arbitrarily small, provided that n and m are sufficiently large, whence for such values of n and m the left hand side is also arbitrarily small, so that, by Cauchy's test, the given series converges. The convergence is absolute, since our argument applies equally well to the convergence of series of absolute values $|a_n|$.

The analogous proof for the following fact will be left to the reader:

If $|a_n| \geqq b_n > 0$ and the series, $\sum_{\nu=1}^{\infty} b_\nu$ diverges, then the series $\sum_{n=1}^{\infty} a_\nu$ is certainly not absolutely convergent. Comparison with the Geometric Series: In applications of the test, the most frequently used comparison series is the geometric series. We obtain immdediately the theorem:

The series $\sum_{n=1}^{\infty} a_n$ is absolutely convergent, if from a certain term onwards there holds a relation of the form,

$$|a_n| < cq^n,$$

where c is a positive number independent of n and q is any fixed positive number less than 1. This test is usually expressed in one of the following, weaker forms: The series,

$$\sum_{n=1}^{\infty} a_n$$

converges absolutely, if there holds from a certain term onwards a relation of the form

$$\left|\frac{a_{n+1}}{a_n}\right| < q,$$

where q is again a positive number less than 1 and independent of n, or, if from a certain term onward there holds a relation of the form

$$\sqrt[n]{|a_n|} < q,$$

where q is again a positive number less than 1. In particular, the conditions of these tests are satisfied if there applies a relation of the form

$$\lim_{n\to\infty}\left|\frac{a_{n+1}}{a_n}\right| = \mathrm{k} < 1$$

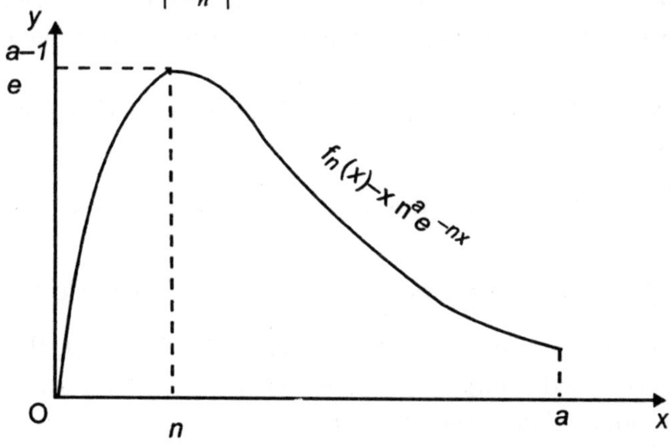

These statements are easily established in the following manner:

Let the criterion II*a*, the ratio test, be satisfied from the suffix

n_0 onward, i.e., when $n > n_0$. For the sake of brevity, let $a_{n_0+m+1} = b_m$ and find that,

$$|b_1| < q|b_0|,\ |b_2| < q|b_1| < q^2|b_0|,\ |b_3| < q, \text{ etc.};$$

hence $|b_m| < q^m|b_0|$, which establi shes our statement. For the Criterion II*b* - the root test -we have at once $|a_n| < q_n$, and our statement follows immediately. We consider an arbitrary number q such that $k < q < 1$. Then, from a certain n_0 onwards, i.e.,, when $n > n_0$, it is certain that,

$$\left.\frac{a_{n+1}}{a_n}\right| < q \text{ or } \sqrt[n]{|a_n|} < q,$$

as the case may be, since from a certain term onwards the values of,

$$\left.\frac{a_{n+1}}{a_n}\right| < q \ of \sqrt[n]{|a_n|}$$

differ from k by less than $(g-k)$. The statement is then established by reference to the results already proved. We stress the point that the four tests derived from the original criterion $|a_n| < cq^n$ are not equivalent to each other or to the original one, i.e., that they cannot be derived from one another in both directions. We shall soon see from examples that, if a series satisfies one of these conditions, it need not by any means satisfy all the other ones. None of these statements can he reversed. For the sake of completeness, it may be pointed out that a series certainly diverges if from a certain term onwards $|a_n| > c$ for a properly chosen positive number c or if from a certain term onwards $\sqrt[n]{|a_n|} > 1$, or if

$$\lim_{n\to\infty}\left|\frac{a_{n+1}}{a_n}\right| = k, \qquad \lim_{n\to\infty}\left|\sqrt[n]{a_n}\right| = k,$$

where $k > 1$. In fact, as we immediately recognize, in such a series the terms cannot tend to zero as n increases, whence the series must diverge. (Under these conditions, the series cannot even be conditionally convergent.) Our tests furnish sufficient conditions for the absolute convergence of a series, i.e., if they are satisfied, we can conclude that the series converges absolutely. However, they are definitely not necessary conditions, i.e., absolutely convergent series can be found which do not satisfy these conditions. For example, the knowledge that,

$$\lim_{n\to\infty}\left|\frac{a_{n+1}}{a_n}\right| = 1, \qquad \lim_{n\to\infty}\left|\sqrt[n]{a_n}\right| = 1,$$

does not entitle us to make any statement about the convergence of the series. Such a series may converge or diverge. For example, the series $\sum_{n=1}^{\infty}\frac{1}{n}$, for which $\lim_{n\to\infty}\sqrt[n]{|a_n|}=1$ and $\lim_{n\to\infty}\sqrt[n]{\left|\frac{a_{n+1}}{a_n}\right|}=1$ is divergent, as we have seen before. On the other hand, we shall soon see that the series $\sum_{n=1}^{\infty}\frac{1}{n^2}$, which satisfies the same relations, converges. As an example of the application of our tests, we first consider the series $q + 2q^2 + 3q^3 +\ldots + nq^n +\ldots$

For this series,

$$\lim_{n\to\infty}\sqrt[n]{|a_n|} = |q|\cdot\lim_{n\to\infty}\sqrt[n]{n} = |q|,$$

$$\lim_{n\to\infty}\left|\frac{a_{n+1}}{a_n}\right| = |q|\cdot\lim_{n\to\infty}\frac{n+1}{n} = |q|.$$

It follows from the ratio test as well as from the root test even in the weaker form III that the series converges, if $|q| < 1$.

On the other hand, if we consider the series $1 + 2q + q^2 + 2q^3 +\ldots + q^{2n} + 2q^{2n+1} +\ldots$, we can no longer prove convergence by the ratio test when ýÿ ≻ $|q| < 1$, because then $|2q^{2n+1}/q^{2n}| = 2|q|$ ≻ 1. But the root test yields immediately $\lim\sqrt[n]{|a_n|} = |q|$ and shows that the series converges provided $|q| < 1$, which, of course, we could have observed directly.

Comparison with an Integral

We now proceed to a discussion of convergence which is independent of the preceding considerations. We shall carry it out for the particularly simple and important case of the series,

$$\sum_{n-1}^{\infty} \frac{1}{n^{\alpha}} = 1 + \frac{1}{2^{\alpha}} + \frac{1}{3^{\alpha}} + \ldots,$$

where the general term a_n is $1/n^a$, $\alpha > 0$. In order to investigate the properties of this series, we consider the graph of the function $y = 1/x\alpha$ and mark off on the x-axis the integral values $x = 1$, $x = 2$. We first construct the rectangle of height $1/n\alpha$ over the interval n-1 ýÿ x ýÿ n of the x-axis ($n > 1$) and compare it with the area of the region bounded by the same interval of the x -axis, the ordinates at the ends and the curve $y=1/x\alpha$. Secondly, we construct the rectangle of height $1/n$a lying above the interval nýxýn+1 and similarly compare it with the area of the region lying above the same interval and below the curve.

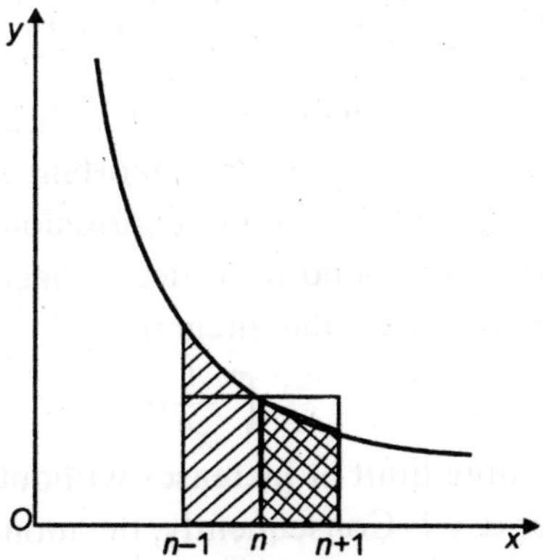

Fig. Comparison of a Series with an Integral

In the first case, the area under the curve is obviously larger than the area of the rectangle; in the second case, it is less than the area of the rectangle.

In other words,

$$\int_{n}^{n+1} \frac{dx}{x^{\alpha}} < \frac{1}{n^{\alpha}} < \int_{n-1}^{n} \frac{dx}{x^{\alpha}},$$

as we may also prove directly from the integral itself. Writing down this inequality for $n = 2$, $n = 3$, …, $n = m$ and summing, we obtain the estimate * for the m-th partial sum,

$$s_m = \sum_{n=1}^{m} \frac{1}{n^{\alpha}}$$

$$1+\int_2^{m+1}\frac{dx}{x^\alpha}<s_m<1+\int_1^m\frac{dx}{x^\alpha}$$

From this relation follows at once that the sequence of numbers,

$$C_n = 1+\frac{1}{2}+\frac{1}{3}+\ldots+\frac{1}{n}-$$

log n is bounded below. Since we see from the inequality,

$$\frac{1}{n+1}<\int_n^{n+1}\frac{dx}{x}=\log(n+1)$$

that the sequence is monotonic decreasing, it must approach a limit,

$$\lim_{n\to\infty} C_n = \lim_{n\to\infty}\left(1+\frac{1}{2}+\frac{1}{3}+\ldots+\frac{1}{n}-\log n\right)$$

The number C with the value 0.577,2 is called *Euler's constant.* In contrast to the other important special numbers of analysis, such as p and e, no other expression with a simple law of formation has been found for Euler's constant.

Now, as m increases, the integral,

$$\int_1^m\frac{1}{x^\alpha}dx$$

tends to a finite limit or increases without limit according to whether $\alpha > l$ or $\alpha > 1$. Consequently, the monotonic sequence of the numbers s_m is bounded or increases beyond all bounds according to whether $\alpha >$ or $\alpha > 1$, whence follows the theorem: The series,

$$\sum_{n=1}^{\infty}\frac{1}{n^\alpha}=\frac{1}{1^\alpha}+\frac{1}{2^\alpha}+\ldots$$

converges - and, of course, absolutely - if and only if, $\alpha > 1$. The divergence of the harmonic series, which we have proved previously in a different way, is an immediate consequence of this. In particular, we see that the series,

$$\frac{1}{1^2}+\frac{1}{2^2}+\frac{1}{3^2}+\ldots,\ \frac{1}{1^3}+\frac{1}{2^3}+\frac{1}{3^3}+\ldots,\ldots \text{ converge.}$$

The series,

$$\sum_{\nu=1}^{\infty} \frac{1}{\nu^{\alpha}},$$

the convergence of which we have just studied, frequently serve as comparison series in investigations of convergence. For example, we see at once that for $\alpha > 1$ the series $\sum_{\nu=1}^{\infty} \frac{c_\nu}{\nu^{\alpha}}$ converges absolutely if the absolute values $|c_n|$ of the coefficients remain less than a fixed bound independent of n.

Sequences and Series of Functions

The terms of the infinite series hitherto considered have been constants, whence these series (when convergent) always represented definite numbers. But both in theory and in applications, the series of outstanding importance are those in which the terms are functions of a variable, so that the sum of the series is also a function of the variable, as in the case of Taylor series.

We shall therefore consider a series $g_1(x) + g_2(x) + g_s(x) + ...$, in which the functions $g_n(x)$ are defined in an interval $a > x > b$. We will denote the n-th partial sum $g_1(x) + g_2(x) + ... + g_n(x)$, by $f_n(x)$. Then the sum $f(x)$ of our series, where it exists, is simply the limit $\lim_{n\to\infty} f_n(x)$.

We may therefore regard the sum of an infinite series of functions as the limit of a sequence of functions $f_1(x)$, $f_2(x)$, $f_n(x)$,... Conversely, for any such sequence of functions $f_1(x)$, $f_2(x)$,..., we can form an equivalent series by setting $g_1(x) = f_1(x)$... and $g_n(x) = f_n(x) - f_{n-1}(x)$ for $n > 1$. Hence, when it is convenient, we can pass from the consideration of series to that of sequences and conversely.

Limiting Processes with Functions and Curves

We shall now state exactly what we mean by saying that a function $f(x)$ is the limit of a sequence $f_1(x), f_2(x), \ldots, f_n(x)$, in an interval $a \rangle x \rangle b$.

The definition is: The sequence $f_1(x), f_2(x),...$ converges in that interval to the limit function $f(x)$, if at each point x of the interval the values $f_n(x)$ converge in the usual sense to the value $f(x)$. In this case, we shall write $\lim_{n\to\infty} f_n(x) = f(x)$.

According to Cauchy's test, we can express the convergence of the sequence without necessarily knowing or stating the limit function $f(x)$. In fact, our sequence of functions will converge to a limit function if, and only if, at each point x in our interval and for every positive number e the quantity $|f_n(x) - f_m(x)|$ is less than e, provided that the numbers n and m are chosen large enough, i.e., larger than a certain number $N = N(\varepsilon)$. This number $N(\varepsilon)$ usually depends on e and x and increases beyond all bounds as ε tends to zero. We have frequently encountered cases of limits of sequences of functions. We mention only the definition of the power x^a for irrational values of α by the equation,

$$x\alpha = \lim_{n\to\infty} x^{r_n}$$

where $r_1, r_2,..., r_n,...$ is a sequence of rational numbers tending to α or the equation,

$$e\alpha = \lim_{n\to\infty}\left(1+\frac{x}{n}\right)^n,$$

where the functions $f_n(x)$ on the right hand side are polynomials of degree n.

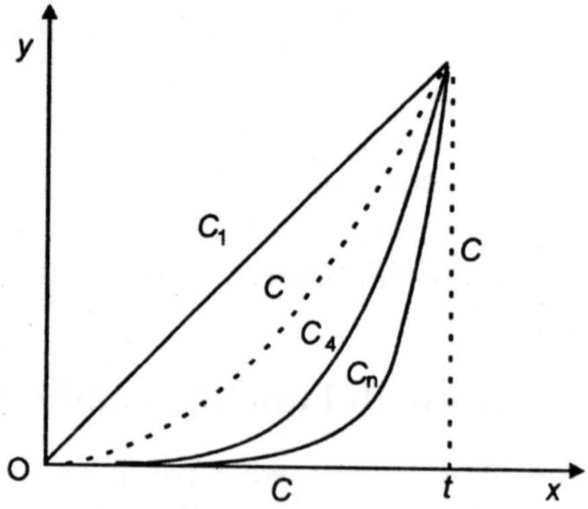

Fig. Limit Curve and Limit Function

The graphical representation of functions by means of curves suggests that we can also speak of limits of sequences of curves, saying, for example, that the graphs of the above limit functions x^a and e^x are to be regarded as the limit curves of the graphs of the functions x^{r_n} *and* $\left(1+\frac{x}{n}\right)^n$ respectively. However, there is a fine distinction between the passages to the limit with functions and with curves. Until the middle of the Nineteenth Century, this distinction was not sufficiently recognized and only by having a clear idea of it can we avoid apparent paradoxes. We shall illustrate this point by an example.

Consider the functions

$$f_n(x) = x^n \ (n = 1, 2, \ldots)$$

in the interval $0 \leq x \leq 1$. All these functions are continuous and the limit function

$$\lim_{n\to\infty} f_n(x) = f(x)$$

exists. However, this limit function is not continuous. On the contrary, since for all values of n the value of the function $f_n(1) = 1$, the limit $f(1) = 1$, while, on the other hand, for $0 \leq x <$ I, the limit $f(x) \lim_{n\to\infty} f_n(x) = 0$. The function $f(x)$ is therefore a discontinuous function which at $x=1$ has the value 1 and for all other values of x in the interval has the value 0.

This discontinuity becomes clear, if we consider the graphs C_n of the functions $y = f_n(x)$. These are continuous curves, all of which pass through the origin and the point $x = 1, y = 1$ and which draw in closer and closer to the x-axis as n increases. The curves possess a limit curve C which is not at all discontinuous, but consists of the portion of the x-axis between $x=0$ and $x=1$ and the portion of the line $x=1$ between $y=0$ and $y=1$. The curves therefore converge to a continuous limit curve with a vertical portion, while the functions converge to a discontinuous limit function.

We thus recognize that this discontinuity of the limit function expresses itself by the occurrence in the limit curve of a portion perpendicular to the x-axis. Such a portion must involve a discontinuity in the limit function and, in fact, such a portion is always present when the limit function is discontinuous. This limit curve is not the graph of the limit function nor can any curve with a vertical portion be the graph of any single-valued function $y = f(x)$; in fact, corresponding to the value of x at which the vertical portion occurs, the curve gives an infinite number of values of y and the function only one. Hence the limit of the graphs of the functions $f_n(x)$ is not the same as the graph of the limit $f(x)$ of these functions.

Uniform and Non-Uniform Convergence

The distinction between the concepts of convergence of functions and of convergence of curves introduces a phenomenon of which it is essential that the student should clearly recognize it. This is the so-called non-uniform convergence of sequences or infinite series of functions. Since it is well known that, as a rule, beginners encounter difficulties here, we should discuss the matter in some detail. By definition, a function $f(x)$ is the limit of a sequence $f_1(x), f_2(x),...$ in an interval $a \succ x \succ b$,only means that the usual limit relationship $f(x) = \lim_{n\to\infty} f(x)$ holds at each point x of the interval. From a naive point of view, one might expect that it would follow automatically from this concept of convergence that if we assign an arbitrary degree of accuracy, say $\varepsilon=1/1000$ or $\varepsilon=1/100$, then from a certain index N onwards all the functions $f_n(x)$ would lie between $f(x) + e$ and $f(x) - \varepsilon$ for all values of x, so that their graphs $y = f_n(x)$ will lie entirely in the strip indicated.

Thus, for every positive ε, there is a corresponding number $N = N(\varepsilon)$, which, of course, will ordinarily increase beyond all bounds as $\varepsilon < 0$, such that, for $n > N$, the difference $|f(x) — f_n(x)| < \varepsilon$, no matter where x is chosen in the interval.

If the accuracy of the approximation can be made at least equal to a pre-assigned number ε everywhere in the interval at the same time, i.e., by everywhere choosing the same number $N(\varepsilon)$ independent of x, we may say that the approximation is uniform. One is at first astonished to find that the naive assumption that convergence is necessarily uniform is altogether wrong; in other words, that convergence may very well be non-uniform.

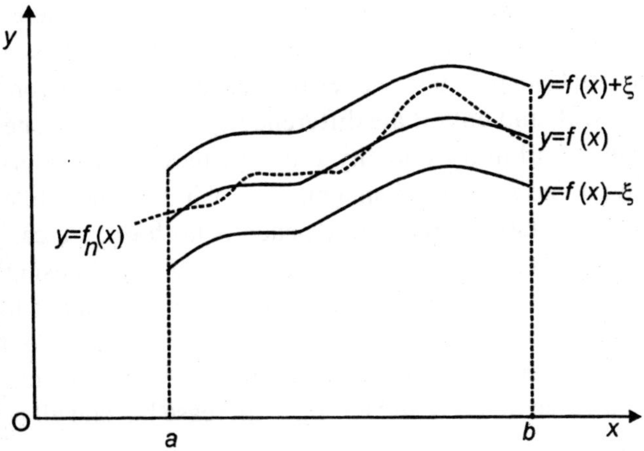

Fig. To Illustrate Uniform Convergence

A Test of Uniform Convergence

The preceding considerations show that the uniform convergence of a sequence or series is a special property which is not possessed by all sequences and series. We shall now reformulate the concept of uniform convergence. The convergent series $g_1(x) + g_2(x) + ...$ is said to be uniformly convergent in an interval, if the sum $f(x)$ can be approximated to within ε (where ε is an arbitrarily small positive number) by taking a number of terms which is sufficiently large and which is the same throughout the interval.

First of all, we assume that the series $g_1(x) + g_2(x) + ...$ converges at every point of a certain interval to a limit function $f(x)$; we denote by $f_n(x)$ the n-th partial sum of the series $f_n(x) = g_1(x) + g_n(x)$ and by $R_n(x)$ the remainder of the series after n terms:

$$R_n(x) = f(x) - f_n(x)$$

The series $g_1(x) + g_2(x) + ...$ is said to be uniformly convergent in the interval if there corresponds to every positive number ε a number N, which only depends on e and not on x, such that for $n > N$ one has the inequality $|R(x)|=|f(x)-f_n(x)|<e$ for all values of x *in* the interval.

Expressed more figuratively, the partial sum $f_n(x)$ represents the sum $f(x)$ to within an error of less than e everywhere in the interval at the same time, provided only that n is chosen large enough. By Cauchy's test, we readily see that the series converges uniformly, if, and only if the difference $|f_n(x) - f_m(x)|$ can be made less than an arbitrary quantity e everywhere in the interval by choosing n and m larger than a number N independent of x.

In fact, firstly, if the convergence is uniform, we can make both $|f_n(x) - f(x)|$ and $|f_m(x) - f(x)|$ less than $e/2$ by choosing n and m larger than a number N independent of x, whence it follows that $|f_n(x) - f_m(x)| < e$; and, secondly, if $|f_n(x) - f_m(x)| < e$, for all values of x whenever n and m are larger than N, then, on choosing any fixed value of $n > N$ and letting m increase beyond all bounds, we have

$$\left|f_n(x) - f(x)\right| = \lim_{m\to\infty}\left|f_n(x) - f_m(x)\right| \leqq \varepsilon,$$

for every value of x, whence the convergence is uniform. If we wish to speak of the uniform convergence of a sequence of functions, we have only to make minute changes in the above definition; the sequence $f_1(x), f_2(x), \ldots$ converges uniformly to $f(x)$ in an interval, if the difference $|f(x) - f_n(x)|$ can be made less than ε everywhere in the interval by choosing n larger than a number N independent of x. As above, a necessary and sufficient condition for the uniform convergence of the sequence is that $|f_n(x) - f_m(x)| < \varepsilon$ for all values of x when both n and m are larger than a certain number N, which depends on ε but not on x.

We shall soon see that it is just this condition of uniform convergence which makes infinite series and other limiting processes with functions into convenient and useful tools of analysis. Fortunately, in the limiting processes, usually encountered in the Calculus and its applications, non-uniform convergence is a sort of exceptional phenomenon which will scarcely trouble us in our present applications of analysis. In most cases, the uniformity of the convergence of a series is established by means of the criterion:

If the terms of the series,

$$\sum_{\nu=1}^{\infty} g_\nu(X)$$

satisfy the condition $\left|g_\nu(X)\right| \leqq a_\nu$, where the numbers a_ϖ are constants which form a convergent series $\sum_{\nu=1}^{\infty} a_\nu$ then the series $\sum_{\nu=1}^{\infty} g_\nu(X)$ converges uniformly (and, we incidentally remark, absolutely).

In fact, we have then,

$$\left|\sum_{\nu=n}^{m} g_\nu(x)\right| \leqq \sum_{\nu=n}^{m} \left|g_\nu(x)\right| \leqq \sum_{\nu=n}^{m} a_\nu,$$

and since, by Cauchy's test, the sum $\sum_{\nu=n}^{m} a_\nu$, can be made arbitrarily small by choosing n and $m > n$ large enough, this expresses exactly the necessary and sufficient condition for uniform convergence. A first example is offered by the geometric series $1 + x + x + \ldots$, where x is restricted to the interval $|x| \leqq q$, a positive number less than 1. The terms of the series are then numerically less than or equal to the terms of the convergent geometric series $\sum q^\nu$.

A further example is the trigonometric series,

$$\frac{c_1 \sin(x-\delta_1)}{1^2} + \frac{c_2 \sin(x-\delta_2)}{2^2} + \frac{c_1 \sin(x-\delta_3)}{3^3} + \ldots,$$

provided that $|c_n| < c$, a positive constant independent of n. Because we have then

$$g_n(x) = \frac{c_n \sin(x-\delta_n)}{n^2}, \text{ so that } \left|g_n(x)\right| < \frac{c}{n^2}$$

Hence the uniform and absolute convergence of the trigonometric series follows from the convergence of the series

Continuity of the Sum of a Uniformly Convergent Series of Continuous Functions

As we have already hinted, the significance of the uniform convergence of an infinite series lies in the fact that a uniformly convergent series in many respects behaves exactly like the sum of a finite number of functions. For example, the sum of a finite number of continuous functions is itself continuous and, correspondingly, we have the theorem: If a series of continuous terms converges uniformly in an interval, its sum is also a continuous function.

The proof is quite simple. We subdivide the series $f(x) = g_1(x) + g_2(x) + ...$ into the m-th partial sum $f_n(x)$ plus the remainder $R_n(x)$. As usual, $f_n(x) = g_1(x) + ... g_n(x)$.

If now any positive number ε is assigned, we can,by virtue of the uniform convergence, choose the number n so large that the remainder is less than $\varepsilon/4$ throughout the entire interval, whence $\left|R_n(x+h) - R_n(x)\right| < \frac{\varepsilon}{2}$ for every pair of numbers x and $x + h$ in the interval. The partial sum $f_n(x)$ consists of the sum of a finite number of continuous functions, whence it is continuous; thus, for each point x in the interval, we can choose a positive δ so small that $\left|f_n(x+h) - f_n(x)\right| < \frac{\varepsilon}{2}$, provided $|h| < \delta$ and the points x and $m + h$ lie in the interval. It then follows that

$$\begin{aligned}\left|f(x+h) - f(x)\right| &= \left|f_n(x+h) - f_n(x) + R_n(x+h) - R_n(x)\right| \\ &\leqq \left|f_n(x+h) - f_n(x)\right| + \left|R_n(x+h) - R_n(x)\right| < \varepsilon,\end{aligned}$$

which expresses the continuity of our function.

The significance of this theorem becomes clear when we recall that the sums of non-uniformly convergent series of continuous functions are not necessarily continuous, as our previous examples show. We may conclude from the preceding theorem that, if the sum of a convergent series of continuous functions has a point of discontinuity, then in every neighbourhood of this point the convergence is non-uniform, whence every representation of discontinuous functions by series of continuous functions is based on the use of non-uniformly convergent limiting processes.

Integration of Uniformly Convergent Series

A sum of a finite number of continuous functions can be integrated term by term, i.e., the integral of the sum can be found by integrating each term separately and adding the integrals. In the case of a convergent infinite series, the same procedure is possible, provided that the series converges uniformly in the interval of integration.

A series $\sum_{\nu=1}^{\infty} g_\nu(X) = f(X)$ which converges uniformly in an interval can be integrated term by term in that interval, or, more precisely, if a and x are two numbers in the interval of uniform convergence, the series $\sum_{\nu=1}^{\infty} \int_a^X g_\nu(t)dt$ converges and, in fact, converges uniformly with respect to x for each fixed value of a, its sum being equal to $\int_a^X f(t)dt$..

In order to prove this we write, as before,

$$f(x) = \sum_{\nu=1}^{\infty} g_\nu(x) = f_n(x) + R_n(x)$$

We have assumed that the separate terms of the series are continuous, the sum is also continuous and therefore integrable. Now, if ε is any positive number, we can find a number N so large that, for every $n > N$, one has $|R_n(x)| < \varepsilon$ for every value of x in the interval. By the first mean value theorem of the integral calculus, we have $\left| \int_a^x \{f(t) - f_n(t)\} dt \right| \leqq \in l$,

where l is the length of the interval of integration. Since the integration of the finite sum $f_n(x)$ can be performed term by term, this yields $\left| \int_a^x f(t)dt - \sum_1^n \int_a^x g_\nu(t)dt \right| \leqq \in l$

However, since $\varepsilon\, l$ can be made as small as we please, this yields $\sum_{\nu=1}^{\infty} \int_a^x g_\nu(t)dt = \lim_{n\to\infty} \sum_{\nu=1}^{n} \int_a^x g_\nu(t)dt = \int_a^x f(t)dt$, as was to be proved.

If we wish to conside instead of infinite series sequences of functions, our result can be expressed as follows:

If in an interval the sequence of functions $f_1(x), f_2(x)$, ... tends uniformly to the limit function $f(x)$, then $\int_a^b f(x)dx = \lim_{n\to\infty} \int_a^b f_n(x)dx$ for every pair of numbers a and b in the interval; in other words, we can then interchange the order of the operations of integration and going to the limit. This fact is far from being a triviality. It is true that, from a naive point of view such as prevailed in the Eighteenth Century, the interchangeabi]ity of the two processes is hardly to be doubted; however, shows that in the case of non-uniform convergence the above equation might not hold. We need only consider Example 2, in which the integral of the limit function is 0, while the integral of the function $f_n(x)$ over the interval $0 \rbrace x \rbrace 1$, i.e., the area of the triangle has the value $\int_a^b f_n(x)\,dx = n^{a-2}$, and, when $\alpha \rbrace 2$, this does not tend to zero. We see here immediately from figure that the reason for the difference between $\int_0^1 f_n(x)\,dx = n^{\alpha-2}$, and $\lim_{n\to\infty} \int_0^1 f_n(x)\,dx$ lies in the non-uniformity of the convergence.

On the other hand, by considering values of α in the interval $1 \rbrace \alpha \rbrace 2$, we see that the equation $\lim_{n\to\infty} \int_0^1 f_n(x)dx = \int_0^1 f(x)dx$ can be valid, although the convergence is non-uniform. As a further example, we can integrate term by term the series $\sum_0^\infty g_n(x)$,, where $g_n(x) = x^n - x^{n-1}$ for $n \rbrace 1$ and $g_0(x) = 1$, between 0 and 1, even though it does not converge uniformly.

Thus, while uniformity of convergence is a sufficient condition for term-by-term integrability, it is by no means a necessary condition. Neglect of this aspect may easily lead to a misunderstanding.

Differentiation of Infinite Series

The behaviour of uniformly converging series or sequences with respect to differentiation is quite different from that with respect to integration.

For example, the sequence of functions $f_n(x) = \dfrac{\sin n^2 \pi}{n}$ certainly converges uniformly to the limit function $f(x) = 0$, but the derivative $f_n'(x) = n \cos n^2 x$ does not converge everywhere to the derivative of the limit function $f'(0) = 0$, as we see by considering $x = 0$. In spite of the uniformity of the convergence, we cannot therefore change the order of the process of differentiation and pass on to the limit.

Naturally, corresponding statements hold for infinite series. For example, the series,

$$\sin x + \frac{\sin 2^4 x}{2^2} + \frac{\sin 3^4 x}{3^2} + \ldots$$

is absolutely and uniformly convergent, because its terms are numerically not greater than the terms of the convergent series

$$\frac{1}{1^2} + \frac{1}{2^2} + \frac{1}{3^2} + \ldots.$$

However, if we differentiate the series term by term, we obtain the series $\cos x + 2^2 \cos 2^4 x + 3^2 \cos 3^4 x + \ldots$, which obviously does not converge everywhere, since, for example, it diverges at $x = 0$. The only useful criterion which assures us in special cases that term-by-term differentiation is permissible is given by the theorem: If, on differentiating a convergent infinite series,

$$\sum_{\nu=0}^{\infty} G_\nu(X) = F(X)$$

term by term, we obtain a uniformly convergent series of continuous terms,

$$\sum_{\nu=0}^{\infty} g_\nu(X) = f(X),$$

then the sum of this last series is equal to the derivative of the sum of the first series. Hence, this theorem expressly requires that, after differentiating the series term by term, we must still investigate whether the result of the differentiation is a uniformly convergent series or not.

The proof of the theorem is almost trivial, because, we can integrate term by term the series obtained by differentiation. Recalling that $g_n(t) = G_n'(t)$, we obtain

$$\int_a^x f(t)\,dt = \int_a^x \left(\sum_{\nu=0}^{\infty} g_\nu(t) \right) dt$$

$$= \sum_{\nu=0}^{\infty} \int_a^x g_\nu(t)\,dt$$

$$= \sum_{\nu=0}^{\infty}\left(G_{\nu}(x)-G_{\nu}(a)\right)$$
$$= F(x) - F(a)$$

Since this is true for every value of x in the interval of uniform convergence, it follows that $f(x) = f(a)$ as was to be proved.

POWER SERIES

Among infinite series, power series occupy the main position. We mean by a power series a series of the type

$$P(x) = c_0 + c_1x + c_2x^2 + ... = \sum_{\nu=0}^{\infty} c_{\nu}x^{\nu}$$

(a power series in x) or, more generally,

$$P(x) = c_0 + c_1x(x - x_0) + c_2(x - x_0)^2 + ... = \sum_{\nu=0}^{\infty} c_{\nu}(x - x_0)^{\nu}$$

(a power series in $(x - x_0)$), where x_0 is a fixed number. If we introduce into the last series $\xi = x - x_0$ as a new variable, it becomes a power series $\sum_{\nu=0}^{\infty} c_{\nu}\xi^{\nu}$ in the new variable ξ, whence, without any loss of generality, we can confine our attention to power series of the more special form $\sum_{\nu=0}^{\infty} c_{\nu}x^{\nu}$

We have considered the approximate representation of functions by polynomials and were thus led to the expansion of functions in Taylor series, which, in fact, are power series. We shall now study power series in somewhat greater detail and obtain expansions of the most important functions in series in simpler and more convenient ways than before.

Convergence Properties of Power Series

There are power series which converge for no value of x, except, of course, for $x = 0$, e.g., the series $x + 2^2x^2 + 3^3x^3 + ... + n^nx^n + ...$ In fact, if $x < 0$, we can find an integer N such that $|x| > 1/N$. Then the absolute values of all the terms n^nx^n, for which $n > N$, will be larger than 1 and, as n increases, nn^nx^n will increase beyond all bounds, so that the series fails to converge. On the other hand, there are series which converge for every value of x; for example, the power series for the exponential function,

$$e^x = 1 + x + \frac{x^2}{2!} + \frac{x^3}{3!} + \ldots,$$

the convergence of which for every value of x follows at once from the ratio test. The $(n + 1)$-th term, divided by the n-th term, yields x/n and, whatever number x is chosen, this ratio tends to zero as n increases. The behaviour of power series with regard to convergence is expressed in the fundamental theorem:

If a power series in x converges for a value $x = \xi$, it converges absolutely for every value x such that $|x| <|\xi|$ and the convergence is uniform in every interval $|x| \leq \eta$, where η is any positive number less than $|\xi|$. Hence h may lie as close to $|\xi|$ as we please.

The proof is simple. If the series $\sum_{\nu=0}^{\infty} c_\nu \xi^\nu$ converges, its terms tend to 0 as n increases. From this follows the weaker statement that all the terms lie below a bound M independent of n, i.e., $|c_n\xi^n| < M$. If now q is any number such that $0<q<1$ and we restrict x to the interval $|x| \leq q|\xi|$, then $|c_n x^n| < |c_n x^n| q^\nu < Mq^\nu$. Hence, in this interval, the terms of the series are smaller in absolute value than the terms of the convergent geometric series ΣMq^ν, whence follows, by the theorem, the absolute and uniform convergence of the series in the interval $-q|\xi| < x < q|x|$.

If a power series does not converge everywhere, i.e., if there is a value $x = x$ for which it diverges, it must diverge for every value of x such that $|x| > |\xi|$. In fact, if it were convergent for such a value of x, by the theorem above, it would have to converge for any numerically smaller value of ξ.

Hence, we recognize that a power series, which converges for at least one value of x other than 0 and which diverges for at least one value of x, has an interval of convergence, i.e., there exists a definite positive number r such that the series diverges for $|x| > \rho$ and converges for $|x| < \rho$. No general statement can be made for $|x| = \rho$. The limiting cases, in which the series converges only for $x=0$ and that which converges everywhere, are expressed symbolically by writing $\rho = 0$ and $\rho = \}$, respectively. It is possible to find this interval of convergence directly from the coefficients c_ν of the series. If the limit $\lim\limits_{n\to\infty} \sqrt[n]{|c_n|}$ exists, then

$$\rho = \frac{1}{\lim\limits_{n\to\infty} \sqrt[n]{|c_n|}}$$

In general, ρ is given by the formula

$$\rho = \frac{1}{\overline{\lim\limits_{n\to\infty}} \sqrt[n]{|c_n|}}$$

where $\overline{\lim}$ is the symbol for the upper limit.

For example, $\rho = 1$ for the geometric series $1 + x + x + \ldots$; at the end-points of the interval of convergence, the series diverges. Similarly, for the series for the inverse tangent

$$\text{arc tan } x = \; - x^3/3 + x^5/5 - + \ldots,$$

we have $\rho = 1$ and at both of the interval's end-points $x = 1$ the series converges, as we recognize at once from Leibnitz's test.

We deduce from the uniform convergence the important fact that within its interval of convergence (if such an interval exists) a power series represents a continuous function. Integration and Differentiation of Power Series: On account of the uniformity of convergence, it is always permissible to integrate a power series

$$f(x) = \sum_{\nu=0}^{\infty} c_\nu x^\nu$$

term by term over any closed interval lying entirely within the interval of convergence.

We thus obtain the function

$$F(x) = c + \sum_{\nu=0}^{\infty} \frac{c_\nu}{\nu+1} x^{\nu+1},$$

for which,

$$F'(x) = f(x)$$

Moreover, since $|c_\nu/(n+1)| \; \} \; |c_\nu|$ for all values of ν, the series obtained by integration converges more rapidly than the original series. We can also differentiate a power series term by term within its interval of convergence, thus obtaining the equation,

$$f'(x) = \sum_{\nu=1}^{\infty} \nu c_\nu x^{\nu-1}$$

In order to prove this statement, we need only show that the series on the right hand side converges uniformly, if x is restricted to an interval lying entirely within the interval of convergence.

Now, let x be a number, lying as close to ρ as we please, for which $\sum_{\nu=1}^{\infty} c_\nu \xi^\nu$ converges; then, as we have seen before, all the numbers $|c_n \xi_n|$ lie below a bound M independently of ν, so that $\left|c_\nu \xi^{\nu-1}\right| < \frac{M}{|\xi|} = N$.

Now let q be any number such that $0 < q < 1$; if we restrict x to the interval $|x| \succ q|\xi|$, the terms of the series under discussion are not larger than those of the series $\sum_{\nu=1}^{\infty} \left|\nu c_\nu q^{\nu-1} \xi^{\nu-1}\right|$, and therefore less than those of the series $\sum_{\nu=1} N\nu q^{\nu-1}$ .However, in this last series, the ratio of the $(n + 1)$-th term to the n-th term is $\{(n+1)/n\}q$, which tends to q as n increases. Since $0<q<1$, it follows by Criterion III*a* that this series converges. Hence the series, obtained by differentiation, converges uniformly and by the theorem at the end of last section represents the derivative $f'_n(x)$ of the function $f(x)$, which proves our statement.

If we apply this result again to the power series

$$f'(x) = \sum_{\nu=1}^{\infty} \nu c_\nu x^{\nu-1},$$

we find on differentiating term by term that

$$f''(x) = \sum_{\nu=2}^{\infty} \nu(\nu-1) c_\nu x^{\nu-2}$$

and, by continuing the process, we arrive at the theorem:

Every function represented by a power series can be differentiated as often as we please within the interval of convergence and the differentiation can be performed term by term.

As an explicit expression for the k-th derivative, we obtain

$$f^{(k)}(x) = \sum_{\nu=k}^{\infty} \nu(\nu-1)\ldots(\nu-k+1)c_\nu x^{\nu-k},$$

or in a slightly different form

$$\frac{f^{(k)}(x)}{k!} = \sum_{\nu=k}^{\infty} \binom{\nu}{k} c_\nu x^{\nu-k} = \sum_{\nu=0}^{\infty} \binom{k+\nu}{k} c_{k+\nu} x^\nu$$

These two formulae are frequently useful.

Operations with Power Series

The preceding theorems on the behaviour of power series are our justification for operating in the same way with power series as with polynomials. Obviously, two power series can be added or subtracted by adding or subtracting the corresponding coefficients.

It is also clear that a power series, like any other convergent series, can be multiplied by a constant factor by multiplying each term by that factor. We merely mention here without a proof that two power series,

$$f(x) = \sum_{\nu=0}^{\infty} a_\nu x^\nu \text{ and } g(x) = \sum_{\nu=0}^{\infty} b_\nu x^\nu$$

can be multiplied by each other like polynomials. In order to be specific, we have the theorem: Throughout the common part of the interval of convergence of two power series, their product is given by the convergent power series $\sum_{\nu=0}^{\infty} c_\nu x^\nu$ with the coefficients

$$\begin{aligned}
c_0 &= a_0 b_0, \\
c_1 &= a_0 b_1 + a_1 b_0, \\
c_2 &= a_0 b_2 + a_1 b_1 + a_2 b_0, \\
&\ldots\ldots\ldots\ldots\ldots\ldots \\
c_n &= a_0 b_n + a_1 b_{n-1} + \ldots + a_n b_0, \\
&\ldots\ldots\ldots\ldots\ldots\ldots
\end{aligned}$$

Theorem of Uniqueness of Power Series: In the theory of power series, the following fact is important: If both power series,

$$\sum_{\nu=0}^{\infty} a_\nu x^\nu \text{ and } \sum_{\nu=0}^{\infty} b_\nu x^\nu$$

converge in an interval which contains the point $x = 0$ inside and if in that interval the two series represent the same function $f(x)$, then they are identical, that is, the equation $a_n = b_n$ is true for every value of n.

In other words: If at all, a function $f(x)$ can be represented by a power series in x in only one way, Briefly speaking, the representation of a function by a power series is unique.

In order to prove this, we need only note that the difference of the two power series, i.e., the power series,

$$\phi(x) = \sum_{\nu=0}^{\infty} c_\nu x^\nu$$

with the coefficients $c_\nu = a_\nu - b_\nu$ represents the function $\phi(x) = f(x) - f(x) = 0$ in the interval, i.e., this last power series converges to the limit 0 everywhere in the interval.

In particular, for $x = 0$, the sum of the series must be 0, i.e., $c_0 = 0$, so that $a_0 = b_0$. We now differentiate the series in the interior of the interval and obtain,

$$\phi'(x) = \sum_{\nu=1}^{\infty} \nu c_\nu x^{\nu-1} .$$

However, $\phi'(x)$ is also 0 throughout the interval, whence, in particular, we have for $x = 0$: $c_1 = 0$ or $a_1 = b_1$. Continuing this process of differentiation and then setting $x = 0$, we find successively that all the coefficients c_n are equal to zero, which proves the theorem. Moreover, we see that we can draw from the above discussion the conclusion: If we take the ν-th derivative of a series, $f(x) = \sum a_\nu x^\nu$ and then set $x = 0$, we at once obtain

$$a_\nu = \frac{1}{\nu!} f^{(\nu)}(0)$$

i.e., Every power series which converges for points other than $x = 0$ is the Taylor series of the function which it represents. The uniqueness of the expansion is here expressed by the fact that the coefficients are uniquely determined by the function itself.

Expansion of Given Functions in Power Series

Within its interval of convergence, every power series represents a continuous function with continuous derivatives of all orders. We shall now discuss the converse problem of the expansion of a given function in a power series. In theory, we can always do this by means of Taylor's theorem, however, in practice, we often encounter difficulties with the actual calculation of the n-th derivative and with the estimation of the remainder. But we can often reach our goal more simply by means of the device:

We first write down the statement,

$$f(x) = \sum_{\nu=0}^{\infty} c_\nu x^\nu,$$

where, to begin with, the coefficients c_ν are unknown. Then we determine by some known property of the function $f(x)$ the coefficients and prove subsequently the convergence of the series. The series represents a function and there remains only to prove that this function is identical to $f(x)$. Due to the uniqueness of the expansion in power series, we know that no other series than the one just found can be the required expansion.

We shall now consider some examples of the use of this method. In fact, we have already obtained the series for artan x and log $(1+x)$ by a method, which forms part of the range of ideas. We simply integrated term by term the series for the derivatives of these functions, which we knew to be geometric series. The Exponential Function: Our problem is to find a function for which $f'(x) = f(x)$ and $f(0) = 1$. If we write down the series with undetermined coefficients $f(x) = c_0 + c_1x + c_2x^2 + \ldots$, and differentiate it, we obtain $f'(x) = c_1 + 2c_2x + 3c_3x^2 + \ldots$.

Since, by assumption, these two power series must be identical, we have the equation $nc_n = c_{n-1}$, which is true for all values of $n > 1$. If we observe that, due to the relation $f(0)=1$, the coefficient c_0 must have the value 1, we can compute all the coefficients successively and obtain the power series $f(x) = 1 + \frac{x}{1!} + \frac{x^2}{2!} + \frac{x^3}{3!} + \ldots$

As is easily shown by the ratio test, this series converges for all values of x and therefore represents a function for which the relations $f'(x) = f(x)$, $f(0) = 1$ are actually fulfilled. Now, the function e^x certainly possesses these properties; we readily deduce that the function $\phi(x)$ is identical to e^x. In fact, if we form the quotient $f(x)=f(x)/e^x$ and differentiate it, we find,

$$\varphi(\xi)=\frac{e^{\alpha} f'(x)-e^{\alpha} f(x)}{e^{2x}}=0$$

Hence, the function $\phi(x)$ is a constant and since it has the value 1 for $x = 0$, it must be identically equal to 1, thus proving that our power series and the exponential function are identical.

The Binomial Series

We can now return to the binomial series, this time using the method of undetermined coefficients. We wish to expand the function $f(x) = (1+x)^{\alpha}$ in a power series, whence we write $f(x) = (1 + x)^{\alpha} = c_0 + c_1x + c_2x^2 + \ldots$, where the coefficients c_n are undetermined. We now note that obviously our function satisfies the relation,

$$(1+x)=f'(x)=\alpha f(x)=\sum_{\nu=0}^{\infty}\alpha c_{\nu}x^{\nu}$$

On the other hand, if we differentiate the series for $f(x)$ term by term and multiply by $(1+x)$, we obtain $(1+x)f'(x)=c_1+(2c_2+c_1)\,x+(3c_3+2c_2)\,x^2+\ldots$; since these two power series for $f'(x)$ must be identical, we have $\alpha c_0 = c_1$, $\alpha c_1 = 2c_2 + c_1$, $\alpha c_2 = 3c_3 + 2c_2,\ldots$.

It is now certain that $c_0 = 1$, since our series must have the value 1 for $x = 0$; we thus find in succession the expressions

$$c_1=\alpha,\ c_1=\frac{(\alpha-1)\alpha}{2},\ c_1=\frac{(\alpha-2)(\alpha-1)\alpha}{3\cdot 2},\ldots,$$ for the coefficients, and, in general, as is readily established,

$$c_n=\frac{(\alpha-\nu-1)(\alpha-\nu+2)\ldots(\alpha-1)\alpha}{\nu(\nu-1)\ldots 2.1}=\binom{\alpha}{\nu}$$

Substituting these values for the coefficients, we have the series $\sum_{\nu=0}^{\infty}\binom{\alpha}{\nu}x^{\nu}$; we must yet investigate the convergence of this series and show that it actually represents $(1+x)^n$.

By the ratio test, we find that, when α is not a positive integer, the series converges if $|x| < 1$ and diverges if $|x| > 1$; in fact, then the ratio of the $(n + 1)$-th term to the n-th term is $\{(\alpha - n + 1)/n\}x$ and the absolute value of this expression tends to $|x|$ as n increases beyond all bounds. We state here without proof the exact conditions under which this series converges. If the index a is an integer, the series terminates and is therefore valid for all values of x (as it becomes the ordinary binomial theorem).

For all other values of α, the series is absolutely convergent for $|x|<1$ and diverges for $|x|>1$. For $x = +1$, the series converges absolutely if $\alpha > 0$, conditionally if $-1 < \alpha < 0$, and it diverges if $\alpha = -1$. Finally, at $x = - 1$, the series is absolutely convergent if $a > 0$, divergent if $a < 0$. Hence, if $|x| < 1$, our series represents a function $f(x)$ which satisfies the condition $(1 + x)f'(x) = af(x)$, as follows from the method of forming the coefficients. Moreover, $f(0) = 1$. But these two conditions ensure that the function $f(x)$ is identical to $(1 + x)\alpha$. In fact, setting

$$\phi(x) = f(x)/(1 + x)^{\alpha},$$

we find that

$$\phi(x) = \frac{(1+x)^{\alpha} f'(x) - \alpha(1+x)^{\alpha-1} f(x)}{(1+x)^{2\alpha}}$$

hence $f(x)$ is a constant and is always equal to 1, since $\phi(0) = 1$. We have thus proved that for $|x| < 1$

$$(1 + \mathrm{x})\alpha = \sum_{\nu=0}^{\infty} \binom{\alpha}{\nu} x^{\nu}$$

which is the binomial series.

We present here the following special cases of the binomial series: The geometric series:

$$\frac{1}{1+x} = (1+x)^{-1} = 1 - x + x^2 - x^3 + x^4 - + \ldots = \sum_{\nu=0}^{\infty} (-1)^{\nu} x^{\nu};$$

the series,

$$\frac{1}{(1+x)^2} = (1 + x)^{-2}$$

$$= 1 - 2x + 3x^2 - 4x^3 + -...$$

$$= \sum_{\nu=0}^{\infty} (-1)^\nu (\nu+1)x^\nu,$$

which may otherwise be obtained from the geometric series by differentiation and the

$$\sqrt{(1+x)} = (1+x)^{\frac{1}{2}}$$

$$= 1 + \frac{1}{2}x - \frac{1}{2.4}x^2 + \frac{1.3}{2.4.6}x^3 - \frac{1.3.5}{2.4.6.8}x^4 + -...,$$

$$\frac{1}{\sqrt{(1+x)}} = (1+x)^{-\frac{1}{2}}$$

$$= 1 + \frac{1}{2}x + \frac{1.3}{2.4}x^2 - \frac{1.3.5}{2.4.6}x^3 \frac{1.3.5.7}{2.4.6.8}x^4 - +...,$$

series the first two or three terms of which are useful expressions.

The Series for arsin x: This series can be established readily by expansion of the expression $1/\sqrt{(1-t^2)}$ as a binomial series

$$(1-t^2)^{-\frac{1}{2}} = 1 + \frac{1}{2}t^2 + \frac{1.3}{2.4}t^4 + ...$$

This series converges if $|t| \not> 1$, whence it converges uniformly for $|t| < q < 1$. On integrating term by term between 0 and x, we obtain

$$\text{arc sin}x = x + \frac{1}{2}\frac{x^3}{3} + \frac{1.3}{2.4}\frac{x^5}{5} + ...;$$

by the ratio test, we find that this series converges if $|x| < 1$ and diverges if $|x| > 1$. The derivation of this series by Taylor's theorem would be decidedly less convenient owing to the difficulty of estimating the remainder.

The Series for arsinh $x = \log\{x + (1 + x)\}$: We obtain this expansion by a similar method. By the binomial theorem, we write down the series for the derivative of arsinh x:

$$\frac{1}{\sqrt{1+x^2}} = 1 - \frac{1}{2}x^2 + \frac{1.3}{2.4}x^4 - \frac{1.3.5}{2.4.6}x^6 + -...,$$

and then integrate term by term to obtain the expansion,

$$\text{arc sinh}x = x - \frac{1}{2}\frac{x^3}{3} + \frac{1.3}{2.4}\frac{x^5}{5} - + \ldots;$$

with the interval of convergence $-1 < x < 1$.

Multiplication of Series

The expansion of the function,

$$\frac{\log(1+x)}{1+x}$$

is a simple example of the application of the rule for the multiplication of power series.

We have only to multiply the logarithmic series log,

$$(1 + x) = x - \frac{x^2}{2} + \frac{x^3}{3} + \frac{x^4}{4} + - \ldots$$

by the geometric series

$$\frac{1}{1+x} = 1 - x + x^2 - x^3 + x^4 - + \ldots;$$

the reader should verify that this yields the remarkable expansion

$$\frac{\log(1+x)}{1+x} = x - \left(1 + \frac{1}{2}\right)x^2 + \left(1 + \frac{1}{2} + \frac{1}{3}\right)x^3$$

$$- \left(1 + \frac{1}{2} + \frac{1}{3} + \frac{1}{4}\right)x^4 + - \ldots$$

for $|x| < 1$.

Example of Term-by-Term Integration (Elliptic Integral):. In previous applications, we have encountered the elliptic integral

$$K = \int_0^{\pi/2} \frac{d\varphi}{\sqrt{\left(1 - k^2 \sin^2 \varphi\right)}} (k^2 < 1)$$

(the period of oscillation of a pendulum. In order to evaluate the integral, we can first expand the integrand by the binomial theorem:

$$\frac{1}{\sqrt{\left(1 - k^2 \sin^2 \varphi\right)}} = 1 + \frac{1}{2}k^2 \sin^2 \varphi + \frac{1.3}{2.4}k^4$$

$$\sin^4 \varphi + \frac{1.3.5}{2.4.6}k^6 \sin^6 \varphi + \ldots,$$

Since $k^2 \sin^2 \varphi$ is never larger than k^2, this series converges uniformly for all values of φ and we may integrate term by term:

$$K = \int_0^{\pi/2} \frac{d\varphi}{\sqrt{1-k^2\sin^2\varphi}}$$

$$= \int_0^{\pi/2} d\varphi + \frac{1}{2}k^2 \int_0^{\pi/2} \sin^2\varphi\, d\varphi + \frac{1.3}{2.4}k^4$$

$$\int_0^{\pi/2} \sin^4\varphi\, d\varphi + \ldots$$

Substitution of their values yields

$$K = \int_0^{\pi/2} \frac{d\varphi}{\sqrt{1-k^2\sin^2\varphi}}$$

$$= \frac{\pi}{2}\left\{1+\left(\frac{1}{2}\right)^2 k^2 + \left(\frac{1.3}{2.4}\right)^2 k^4 + \left(\frac{1.3.5}{2.4.6}\right)k^6 + \ldots\right\}$$

POWER SERIES WITH COMPLEX TERMS

Complex Terms into Power Series

The similarity between certain power series representing functions which are apparently unrelated led Euler to set up a purely formal link between them, found by giving complex values, in particular, pure imaginary values to the variable x. We shall first do this formally, unhindered by questions of rigour and investigate the results of the process.

The first striking relation of this sort is obtained, if we replace the quantity x in the series for e^x by a pure imaginary $i\phi$, where ϕ is a real number. If we recall the fundamental equation for the imaginary unit i, that is, $i = -1$, from which it follows that $i^3 = -i$, $i^4 = 1$, $i^5 = i, \ldots$, then, on separating the real and imaginary terms of the series, we obtain

$$e^1\phi = \left(1 - \frac{\phi^2}{2!} + \frac{\phi^4}{4!} + \frac{\phi^6}{6!} + - \ldots\right) + i\left(\phi - \frac{\phi^3}{3!} + \frac{\phi^5}{5!} - \frac{\phi^7}{7!} + - \ldots\right),$$

or, in another form, $e^1\phi = \cos\phi + i \sin\phi$

This is the well-known and important Euler formula. which

is as yet purely formal. It is consistent with De Moivre's theorem which is expressed by the equation $(\cos\phi + i \sin\phi)(\cos\psi + i \sin\psi) = (\cos\phi + \psi) + i \sin(\phi + \psi)$. By virtue of Euler's formula, this equation merely states that the relation $e^{x} \cdot e^{v} = e^{x+v}$ continues to hold for pure imaginary values $x = i\phi$, $y = i\psi$.

If we replace the variable x in the power series for $\cos x$ by the pure imaginary ix, we obtain at once the series for $\cosh x$; this relation can be expressed by the equation

$$\cosh x = \cos ix.$$

In the same way, we obtain

$$\sinh x = \frac{1}{i}\sinh ix.$$

Since Euler's formula also yields

$$e^{-if} = \cos\phi - i \sin\phi,$$

we arrive at the exponential expressions for the trigonometric functions

$$\sin x = \frac{e^{ix} - e^{-ix}}{2i}, \qquad \cos x = \frac{e^{ix} + e^{-ix}}{2}$$

These are exactly analogous to the exponential expressions for the hyperbolic functions and are, in fact, transformed into them by the relations

$$\cosh x = \cos ix, \qquad \sinh x = \frac{1}{i}\sin ix$$

Naturally, corresponding formal relations can also be obtained for the functions tan x, tanh x, cot x and coth x, which are interrelated by the equations

$$\tanh x = \frac{1}{i}\tan ix \qquad \coth x = i \cot ix$$

Finally, similar relations can also be found for the inverse trigonometric and hyperbolic functions. For example, we immediately find from

$$y = \tan x = \frac{e^{ix} - e^{-ix}}{i\left(e^{ix} + e^{-ix}\right)} = \frac{e^{2ix} - 1}{i\left(e^{2ix} + 1\right)}$$

that

$$e^{2ix} = \frac{1 + iy}{1 - iv}$$

It we take the logarithms of both sides of this equation and then write x instead of y and artan x instead of x, we obtain the equation,

$$\text{artan}x = \frac{1}{2i}\log\frac{1+ix}{1-ix},$$

which expresses a remarkable connection between the inverse tangent and the logarithm. If we replace in the known power series for $\log(1 + x)/(1 - x)$ x by ix, we actually obtain the power series for artan x:

$$\text{artan}x = \frac{1}{i}\left(ix + \frac{(ix)^3}{3} + \frac{(ix)^5}{5} + ...\right) = x - \frac{x^3}{3} + \frac{x^5}{5} - +...$$

The above relations have as yet a purely formal character and naturally demand a more exact statement as to the meaning which they are intended to convey. In the next sub-section, we shall indicate how this can be given with the help of function theory.

However, we shall later on only require Euler's formula $e^{i\phi} = \cos\phi + i\sin\phi$ This being so, we can avoid here a thorough analysis. We need only regard the symbol $e^i\phi$ as a formal abbreviation $\cos\phi + i\sin\phi$, for in which case De Moivre's formula $e^{i\phi} \cdot e^{i\psi} = e^{i(\phi+\psi)}$ appears merely as a consequence of the elementary addition theorem of trigonometry.

From this formal point of view, in order to make the relation $e^x \cdot e^y = e^{x+y}$ e*. e* remain valid for any complex arguments, we set up the further definition

$$e^x = e^{\xi}(\cos\eta + i\ \text{xin}\eta)$$

where,

$$x = \xi + i\eta\ (\xi, \eta\ \text{real}).$$

4

Continuous Functions

A continuous function is a function for which, intuitively, small changes in the input result in small changes in the output. Otherwise, a function is said to be discontinuous. A continuous function with a continuous inverse function is called bicontinuous. An intuitive though imprecise (and inexact) idea of continuity is given by the common statement that a continuous function is a function whose graph can be drawn without lifting the chalk from the blackboard.

Continuity of functions is one of the core concepts of topology, which is treated in full generality in a more advanced article. This introductory article focuses on the special case where the inputs and outputs of functions are real numbers. In addition, this article discusses the definition for the more general case of functions between two metric spaces. In order theory, especially in domain theory, one considers a notion of continuity known as Scott continuity.

As an example, consider the function $h(t)$ which describes the height of a growing flower at time t. This function is continuous. In fact, there is a dictum of classical physics which states that in nature everything is continuous. By contrast, if $M(t)$ denotes the amount of money in a bank account at time t, then the function jumps whenever money is deposited or withdrawn, so the function $M(t)$ is discontinuous.

IDEA OF LIMIT

We thought of the graph of a function, which is continuous in an interval, as a curve consisting of one unbroken piece; we also stated that the change in the function y must remain arbitrarily small, provided only that the change of the independent variable x is restricted to a sufficiently small interval. This state of affairs is usually formulated as follows, with a greater range but increased precision. A function $f(x)$ is said to be continuous at the point ξ, if it possesses the property: At the point ξ, the value of the function $f(\xi)$ is approximated to within an arbitrary pre-assigned degree of accuracy ε by all functional values $f(x)$ for which x is near enough to ξ. In other words, $f(x)$ is continuous at ξ, if for every positive number ε, no matter how small, there can be determined another positive number $\delta = \delta(\varepsilon)$ such that $|f(x) - f(\xi)| < \varepsilon$ for all points x for which,

$$|x - \xi| < \delta.$$

Or again: The condition of continuity requires that for the point ξ

$$\lim_{x\to\xi} f(x) = f(\xi)$$

The value of the function at the point ξ is the same as the limit of the functional values $f(x_n)$ for any arbitrary sequence x_n of numbers converging to ξ.

It is important to observe that our condition involves two different matters:

1. The existence of $\lim_{x\to\xi} f(x)$ and
2. The coincidence of this limit with $f(x)$, the value of the function at the point x.

Having now defined continuity of a function $f(x)$ at a point ξ, we proceed to state what we mean by the continuity of a function $f(x)$ in an interval. This may be defined simply as follows: The function $f(x)$ is continuous in an interval, if it is continuous at each point of that interval. Stated fully, this requires that, if a positive number ε be assigned, then there exists for each point x of the interval a number δ, depending as a rule on ε and on x, such that

$$|f(\overline{x}) - f(x)| < \in \text{ if } |\overline{x} - x| < \delta$$

and $\overline{x}$ lies in the interval $a < \overline{x} < b$.

Closely related to this is another concept, namely that of uniform continuity. A function $f(x)$ is uniformly continuous in the interval $a \wr x \wr$b, if there exists for every positive number ε a corresponding positive number δ such that, for every pair of points x_1, x_2 in the interval, the distance $|x_1 - x_2|$ a part of which is less than δ, one has the inequality $|f(x_1) - f(x_2)| < \varepsilon$.

This differs from the definition stated above in that δ in the definition of uniform continuity does not depend on x, but is equally effective for all values of x, whence follows the term uniform continuity. It is quite obvious that a uniformly continuous function is necessarily continuous. Conversely, it can be shown that every function $f(x)$, which is continuous in a closed interval $a\}x\}b$, is also uniformly continuous.

DISCONTINUOUS FUNCTION

The simplest discontinuity occurs at those points at which a function has a jump, that is, at which the function has a definite limit as x tends to the point from the right and a definite limit as x tends to the point from the left, while these two limits are different. It does not matter whether or how the function is defined at the point of discontinuity itself.

For example, the function $f(x)$ defined by the equations

$$f(x) = 0 \text{ for } x^2 > 1, f(x) = 1 \text{ for}$$

$$x^2 > 1, f(x) = \frac{1}{2} \text{ for } x^3 = 1$$

has discontinuities at the points $\xi = 1$ and $\xi = -1$. The limits on approaching these points from the right and from the left hand side differ by 1, and the values of the function at these points agree with neither limit, but are equal to the arithmetic mean of the two limits. It may be noted in passing that our function can be represented, using the idea of a limit, by the expression

$$f(x) = \lim_{x \to \infty} \frac{1}{1 + x^{2n}}$$

For, if $x < 1$, that is, if x lies in the interval $-1 < < 1$, the numbers x^{2n} will have the limit 0, and the function will have the value 1. However, if $x > 1$, as n increases, x^{2n} will increase beyond all bounds; our function will then have the value 0.

Finally, obviously, for $x^2 = 1$, that is, for $x = +1$ and $x = -1$, the value of the function is.

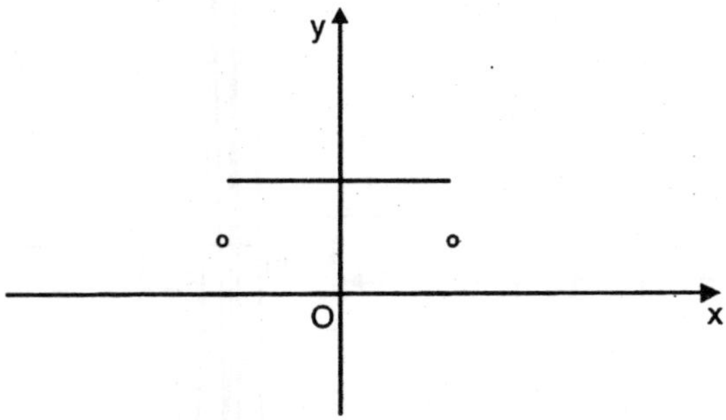

Other curves with jumps are shown in Figures; they represent functions with obvious discontinuities.

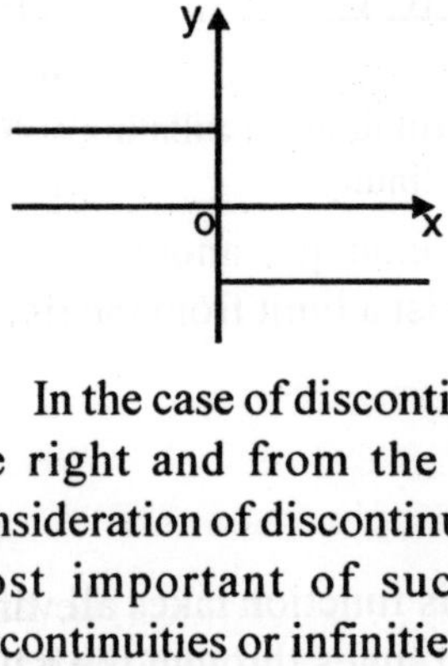

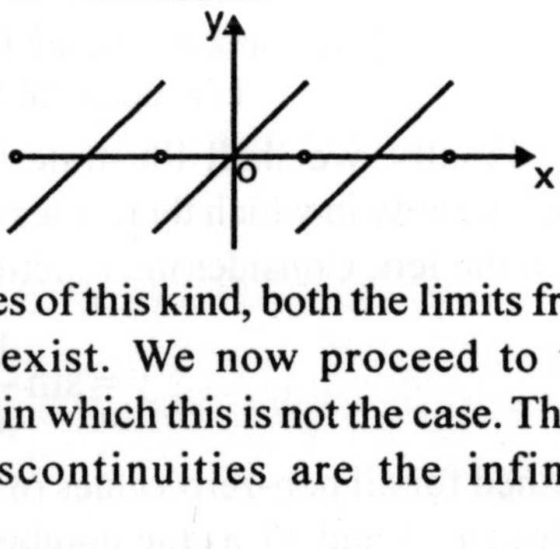

In the case of discontinuities of this kind, both the limits from the right and from the left exist. We now proceed to the consideration of discontinuities in which this is not the case. These most important of such discontinuities are the infinite discontinuities or infinities.

These are discontinuities which are exhibited by the functions $1/x$ or l/x at the point $\xi = 0$; as x ξ, the absolute value $|f(x)|$ of the function increases beyond all bounds.

In the case of $1/x$, the function increases numerically beyond all bounds through positive and through negative values, respectively, as x approaches the origin from the right and from the left.

On the other hand, the function l/x has for $x = 0$ an infinite discontinuity at which its value becomes positive from both sides. The function $y = 1/(x - 1)$, shown in Figure, has infinite discontinuities both at $x = 1$ and at $x = -1$.

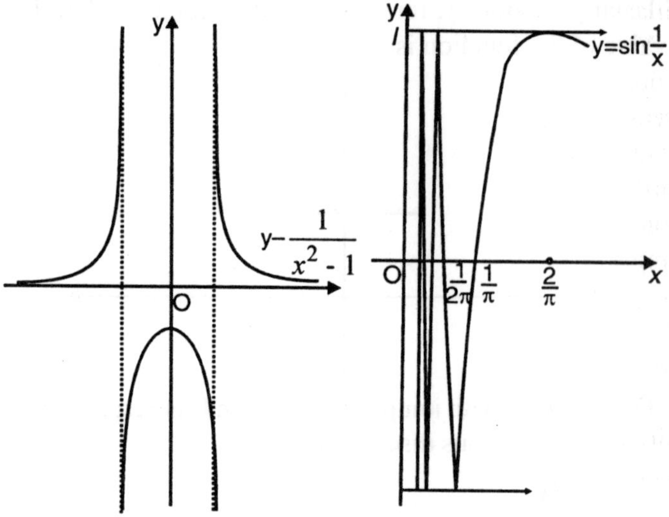

Fig. Function with Infinite Discontinuities, Oscillating Function with Discontinuity

Finally, we shall illustrate by an example another type of discontinuity in which there does not exist a limit from the right or from the left. Consider the function

$$y = \sin\frac{1}{x}$$

defined for all non-zero values of x. This function takes all values between -1 and $+1$ as the number $1/x$ ranges through the values from $(2n - 1/2)\pi$ to $(2n+1/2)\pi$, no matter what value has n. At the points $x=2/(4n-1)\pi$, the function will have the value -1, at the points $x = 2/(4n+1)\pi$ the value $+1$.

We see from this that the function swings backwards and forwards more rapidly between the values $+1$ and -1 as x approaches closer and closer to the point 0 and that there occur in the immediate neighbourhood of the point $x = 0$ an infinite number of oscillations. It is interesting to observe that, in contrast to the above example, the function $y = x\sin x/x$ remains continuous at the point $x = 0$, if we assign to it the value 0 at that point. This continuity is due to the fact that, as the origin is approached, the factor x damps the oscillations of the sine.

Yet, in the neighbourhood of the origin, the function $y = x \sin 1/x$ does not change a finite number of times from monotonic increasing to monotonic decreasing.

On the contrary, it oscillates backwards and forwards an infinite number of times, the magnitude of these oscillations becoming as small as we please as the origin is approached. This example shows us that even the simple idea of continuity admits all sorts of remarkable possibilities foreign to our naive intuition. There is one important fact which must be taken into account, if we are to give our ideas greater precision.

It may happen that at a certain point a function is not defined by the original law, as for example at the point $x = 0$ in the last two examples discussed. We have then the right to extend the definition of the function by assigning to it any desired value at such a point. However, in the last example, we can extend the definition in such a way that the function also remains continuous at that point, namely, by setting $y = 0$ when $x = 0$.

This can be done whenever both the limits from the left and from the right exist and are equal to each other; we need then only make the value of the function at the point in question equal to these limits, in order to make the function continuous there. In the case of the function $y = \sin 1/a$, this cannot be done.

Continuous Functions Theorem

In conclusion, we quote the following important, general theorems, the proofs of which follow immediately from the remarks on operations with limits:

The sum, difference, and product of two continuous functions are themselves continuous. The quotient of two continuous functions is continuous at every point at which the denominator does not vanish. In particular, it follows that all polynomials and all rational functions are continuous except at the points where the denominator vanishes. The fact that the other elementary functions, such as the trigonometric functions, are continuous will follow naturally from later considerations.

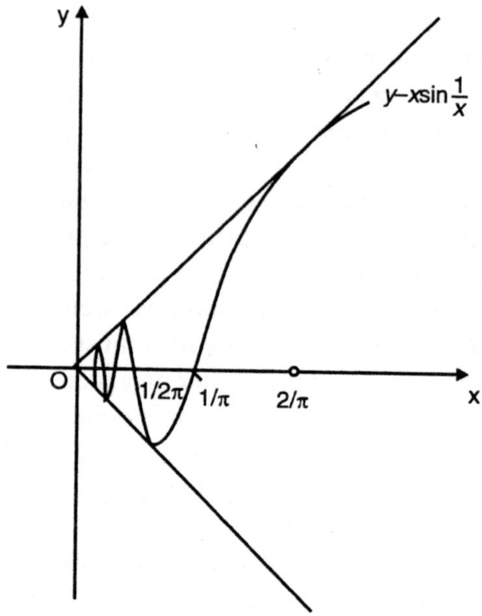

ONE SIDED LIMITS

Although sometimes we get results directly, it is usually helpful to have a larger range of techniques. We give one here and more in section.

***Definition*:** Say that $\lim\limits_{x \to a-} f(x) = l$, or that f has a limit from the left iff given $\in > 0$, there is some $\delta > 0$ such that whenever $a - \delta < x < a$, then $|f(x) - f(a)| < \in$. There is a similar definition of "limit from the right", writen as $\lim\limits_{x \to a+} f(x) = l$

Example: Define $f(x)$ as follows:

$$f(x) = \begin{cases} 3 - x & \text{if } x < 2 \\ 2 & \text{if } x < 2; \\ x/2 & \text{if } x > 2 \end{cases}$$

Calculate the left and right hand limits of $f(x)$ at 2.

Solution: As $x \to 2-$, $f(x) = 3 - x \to 1+$, so the left hand limit is 1. As $x \to 2+$, $f(x) = x/2 \to 1+$, so the right hand limit is 1. Thus the left and right hand limits agree (and disagree with $f(2)$, so f is not continuous at 2). Note our convention: If $f(x) \to 1$ and always $f(x) \geq 1$ as $x \to 2-$, we say that $f(x)$ tends to 1 from above, and write $f(x) \to 1+$ etc.

Proposition: If $\lim_{x \to a} f(x)$ exists, then both one sided limts exist and are equal. Conversely, if both one sided limits exits and are equal, then $\lim_{x \to a} f(x)$ exists. This splits the problem of finding whether a limit exists into two parts; we can look on either side, and check first that we have a limit, and second, that we get the same answer. For example both 1-sided limits exist, but are not equal. There is now an obvious way of checking continuity.

Continuity Test

Proposition: The function f is continuous at a iff both one sided limits exits and are equal to $f(a)$.

Example: Let $f(x) = \begin{cases} x^2 \text{ for } x \leq 1, \\ x \text{ for } x \geq 1. \end{cases}$

Show that f is continuous at 1.

Solution: We use the above criterion. Note that $f(1) = 1$. Also

$$\lim_{x \to 1-} f(x) = \lim_{x \to 1-} x^2 = 1$$

while $$\lim_{x \to 1+} f(x) = \lim_{x \to 1+} x = 1 = f(1).$$

so f is continuous at 1.

Example: Let $f(x) = |x|$. Then f is continuous in $\mathbb{R}$.

Solution: Note that if $x < 0$ then $|x| = -x$ and so is continuous, while if $x > 0$, then $|x| = x$ and so also is continuous. It remains to examine the function at 0. From these identifications, we see that,

$$\lim_{x \to 0-} |x| = 0+,$$

while $\lim_{x \to 0+} |x| = 0+$. Since $0+ = 0- = 0 = |0|$, by the $|x|$ is continuous at 0.

RESULTS GIVING CONTINUITY

Just as for sequences, building continuity directly by calculating limits soon becomes hard work. Instead we give a result that enables us to build new continuous functions from old ones, just as we did for sequences.

Note that if f and g are functions and k is a constant, then $k.f$, $f + g$, fg and (often) f/g are also functions.

Proposition: Let f and g be continuous at a, and let k be a constant. Then $k.f$, $f + g$ and fg are continuous at f. Also, if $g(a) \neq 0$, then f/g is continuous at a.

Proof: We show that $f + g$ is continuous at a. Since, by definition, we have $(f + g)(a) = f(a) + g(a)$, it is enough to show that

$$\lim_{x \to \alpha} \left(f(x) + f(x)\right) 4 = f(a) + g(a).$$

Pick $\in > 0$; then there is some δ_1 such that if $|x - a| < \delta_1$, then $|f(x) - f(a)| </2$. Similarly there is some δ_2 such that if $|x - a| < \delta_2$, then $|g(x) - g(a)| < \in/2$. Let $= \min(\delta_1, \delta_2)$, and pick x with $|x - a| < \delta$. Then $|f(x) + g(x) - (f(a) + g(a))| \leq |f(x) - f(a)| + |g(x) - g(a)| < \in/2 + \in/2 = \in$.

This gives the result. The other results are similar, but rather harder; for proofs.

Note: Just as when dealing with sequences, we need to know that f/g is defined in some neighbourhood of a. This can be shown using a very similar proof to the corresponding result for sequences.

Proposition: Let f be continuous at a, and let g be continuous at $f(a)$. Then gof is continuous at a

Proof: Pick $\in > 0$. We must find $\delta > 0$ such that if $|x - a| < \delta$, then $g(f(x)) - g(f(a))| < \in$. We find δ using the given properties of f and g. Since g is continuous at $f(a)$, there is some $\delta_1 > 0$ such that if $|y - f(a)| < \delta_1$, then $|g(y) - g(f(a))| < \in$.

Now use the fact that f is continuous at a, so there is some $\delta > 0$ such that if $|x - a| < \delta$, then $|f(x) - f(a)| < \delta_1$. Combining these results gives the required inequality.

INFINITE LIMITS

There are many more definitions and results about limits. First one that is close to the sequence definition:

Definition: Say that $\lim_{x\to\infty} f(x) = l$ iff given $\in > 0$, there is some K such that whenever $x > K$, then $|f(x) - l| < \in$.

Example: Evaluate $\lim_{x\to\infty} \dfrac{x^2+3}{3x^2+2x+1}$.

Solution: The idea here should be quite familiar from our sequence work. We use the fact that $1/x \to 0$ as $x \to \infty$. Thus

$$\frac{x^2+3}{3x^2+2x+1} = \frac{1+3/x^2}{3+2/x+1/x^2} \to \frac{1}{3} \text{ as } x \to \infty.$$

Definition Say that $\lim_{x\to\infty} f(x) = \infty$ iff given $L > 0$, there is some K such that whenever $x > K$, then $f(x) > L$.

The reason for working on proofs from the definition is to be able to check what results of this type are trivially true without having to find it in a book. For example

Proposition: Let $g(x) = 1/f(x)$. Then $g(x) \to 0+$ as x iff $f(x)$ as x. Let $y = 1/x$. Then $y \to 0+$ as $x \to \infty$; conversely, $y \to \infty$ as $x \to 0+$

Proof: Pick $\in > 0$. We show there is some K such that if $x > K$, then $0 < y < \in$; indeed, simply take $K = 1/\in$. The converse is equally trivial.

CLOSED INTERVAL CONTINUITY

So far our results have followed because f is continuous at a particular point. In fact we get the best results from assuming rather more. Indeed the results in this section are precisely why we are interested in discussing continuity in the first place. Although some of the results are "obvious", they only follow from the continuity property, and indeed we present counterexamples whenever that fails. So in order to be able to use the following helpful results, we must first be able to check our functions satisfy the hypothesis of continuity. That task is of course what we have just been concentrating on.

Definition: Say that f is continuous on $[a, b]$ iff f is continuous on (a, b), and if, in addition, $\in \lim_{x \to a+} f(x) = f(a)$, and $\in \lim_{x \to b-} f(x) = f(b)$. We sometimes refer to f as being continuous on a compact interval. Such an f has some very nice properties.

Theorems on Boundedness

Theorem Let f be continuous on the compact interval $[a, b]$. Then there are constants M and m such that $m < f(x) < M$ for all $x \in [a, b]$. In other words, we are guaranteed that the graph of f is bounded both above and below.

Proof: This again uses the completeness of $\mathbb{R}$, and again no proof is offered. Note also that the hypotheses are all needed.

Theorem: Let f be continuous on the compact interval $[a, b]$. Then there are points x_0 and x_1 such that $f(x_0) < f(x) < f(x_1)$ for all $x \in [a, b]$.

In other words, we are guaranteed that the graph of f is bounded both above and below, and that these global extrema are actually attained.

Proof: This uses the completeness of $\mathbb{R}$, and follows in part from the previous result.

Consider the function $g(x) = (M - f(x))^{-1}$. This is clearly continuous. If there is some point at which $f(y) = M$, there is nothing to prove; otherwise $g(x)$ is defined everywhere, and is continuous, and hence bounded. This contradicts the fact that M was a least upper bound for f.

We have seen that any polynomial function $P(x)$ satisfies:

$$\lim_{x \to a} P(x) = P(a)$$

for all real numbers a. This property is known as continuity.

Definition: Let $f(x)$ be a function defined on an interval around a. We say that $f(x)$ is continuous at a iff

$$\lim_{x \to a} f(x) = f(a)$$

Otherwise, we say that $f(x)$ is discontinuous at a.

Note that the continuity of $f(x)$ at a means two things:

1. $\lim_{x \to a} f(x)$ exists,
2. And this limit is $f(a)$.

So to be discontinuous at a, means:

1. $\lim_{x \to a} f(x)$ does not exist,
2. or if $\lim_{x \to a} f(x)$ exists, then this limit is not equal to $f(a)$.

BASIC PROPERTIES OF LIMITS

Theorem. If $f(x)$ and $g(x)$ are continuous at a. Then

1. $f(x) + g(x)$ is continuous at a;
2. $\alpha f(x)$ is continuous at a, where is an arbitrary number;
3. $f(x)g(x)$ is continuous at a;
4. $\dfrac{f(x)}{g(x)}$ is continuous at a, provided $g(a) \neq 0$;
5. If $f(x)$ is positive, i.e. $f(x) \geq 0$, then $\sqrt{f(x)}$ is continuous at a;
6. If $f(x)$ is continuous at a and $g(x)$ is continuous at $f(a)$, then their composition $g \circ f(x)$ is continuous at a.

Remark: Many functions are not defined on open intervals. In this case, we can talk about one-sided continuity. Indeed, $f(x)$ is said to be continuous from the left at a iff

$$\lim_{x \to a-} f(x) = f(a)$$

and $f(x)$ is said to be continuous from the right at a iff

$$\lim_{x \to a+} f(x) = f(a)$$

Example: The function $f(x) = \sqrt{x}$ is defined for $x \geq 0$. So we can not talk about left-continuity of $f(x)$ at 0. But since we conclude that $f(x)$ is right-continuous at 0. This concept is also important for step-functions.

Example: Consider the function

$$f(x) = \begin{cases} x^3 + 2 & if \quad x < 2 \\ 5 & if \quad x = 2 \\ x^2 + 6 & if \quad x > 2 \end{cases}$$

The details are left to the reader to see

$$\lim_{x \to 2-} f(x) = \lim_{x \to 2-} x^3 + 2 = 10,$$

and

$$\lim_{x \to 2+} f(x) = \lim_{x \to 2+} x^2 + 6 = 10.$$

So we have

$$\lim_{x \to 2} f(x) = 10.$$

Since $f(2) = 5$, then $f(x)$ is not continuous at 2.

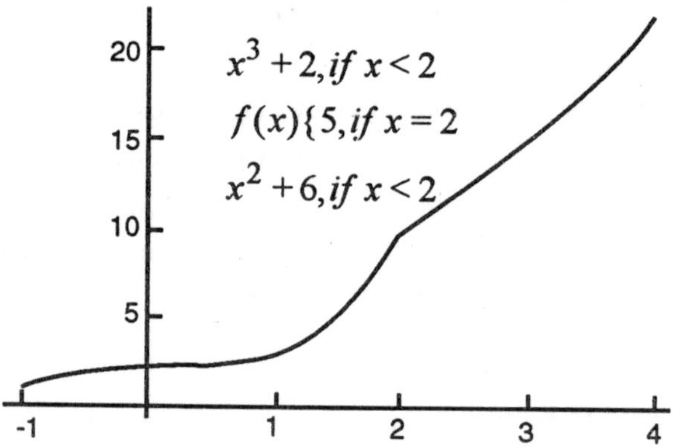

THE INTERMEDIATE VALUE THEOREM

We can draw the graph of a continuous function without lifting your pencil off the paper. This is made precise by the following result: Intermediate Value Theorem. Let $f(x)$ be a continuous function on the interval $[a, b]$. If $d \in [f(a), f(b)]$, then there is a $c \in [a, b]$ such that $f(c) = d$. In the case where $f(a) > f(b)$, $[f(a), f(b)]$ is meant to be the same as $[f(b), f(a)]$. Another way to state the Intermediate Value Theorem is to say that the image of a closed interval under a continuous function is a closed interval. We will present an outline of the proof of the Intermediate Value Theorem.

Here is a classical consequence of the Intermediate Value Theorem:

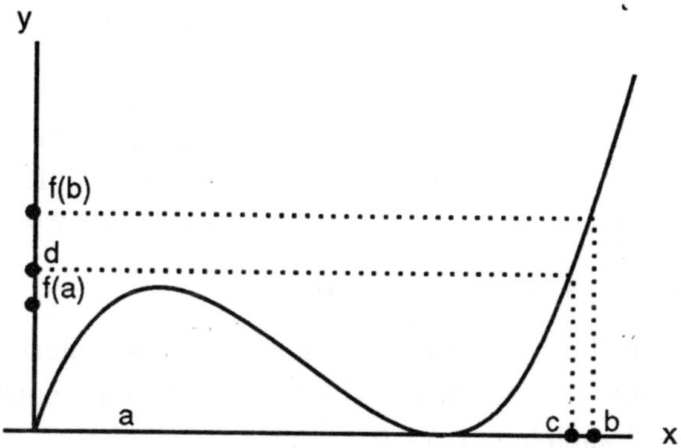

Example: Every polynomial of odd degree has at least one real root. We want to show that if $P(x) = a_n x^n + a_{n-1} x^{n-1} + \dots + a_1 x + a_0$ is a polynomial with n odd and $a_n \neq 0$, then there is a real number c, such that $P(c) = 0$.

First let me remind you that it follows from the results in previous pages that every polynomial is continuous on the real line. There you also learned that

$$\lim_{x \to \infty} \frac{P(x)}{a_n x^n} = 1 \text{ and } \lim_{x \to -\infty} \frac{P(x)}{a_n x^n} = 1$$

Consequently for $|x|$ large enough, $P(x)$ and $a_n x^n$ have the same sign. But $a_n x^n$ has opposite signs for positive x and negative x. Thus it follows that if $a_n > 0$, there are real numbers $x_0 < x_1$ such that $P(x_0) < 0$ and $P(x_1) > 0$.

Similarly if $a_n < 0$, we can find $x_0 < x_1$ such that $P(x_0) > 0$ and $P(x_1) < 0$. In either case, it now follows directly from the Intermediate Value Theorem that (for $d = 0$) there is a real number $c \in [x_0, x_1]$ with $P(c) = 0$.

The natural question arises whether every function which satisfies the conclusion of the Intermediate Value Theorem must be continuous. Unfortunately, the answer is no and counterexamples are quite messy. The easiest counterexample is the function

$$f(x) = \begin{cases} \sin\left(\dfrac{1}{x}\right) & ,\text{if } x \neq 0 \\ 0 & ,\text{if } x \neq 0 \end{cases}$$

This function fails to be continuous at $x = 0$. On the other hand, it is not too hard to see that $f(x)$ has the "Intermediate Value Property" even on closed intervals containing $x = 0$.

Intermediate Value Theorem

Theorem: Let f be continuous on the compact interval $[a, b]$, and assume that $f(a) < 0$ and $f(b) > 0$. Then there is some point c with $a < c < b$ such hat $f(c) = 0$.

Proof: We make no attempt to prove this formally, but sketch a proof with a pair of sequences and a repeated bisection argument. It is also noted that each hypothesis is necessary.

Example: Show there is at least one root of the equation $x - \cos x = 0$ in the interval $[0, 1]$.

Proof: Apply the Intermediate Value Theorem to f on the closed interval $[0, 1]$. The function is continuous on that interval, and $f(0) = -1$, while $f(1) = 1 - \cos(1) > 0$. Thus there is some point $c \in (0, 1)$ such that $f(c) = 0$ as required.

Exercise: Show there is at least one root of the equation $x - e^{-x} = 0$ in the interval $(0, 1)$.

Corollary: Let f be continuous on the compact interval $[a, b]$, and assume there is some constant h such that $f(a) < h$ and $f(b) > h$. Then there is a point c with $a < c < b$ such that

$f(c) = h$.

Proof: Apply the Intermediate Value Theorem to $f - h$ on the closed interval [0, 1]. Note that by considering $-f + h$ we can cope with the case when $f(a) > h$ and $f(b) < h$.

Note: This theorem is the reason why continuity is often loosely described as a function you can draw without taking your pen from the paper. As we have seen with $y = 1/x$, this can give an inaccurate idea; it is in fact more akin to connectedness.

THE BISECTION METHOD

The Bisection Method at the same time gives a proof of the Intermediate Value Theorem and provides a practical method to find roots of equations. If your calculator can solve equations numerically, it most likely uses a combination of the Bisection Method and the Newton-Raphson Method.

Recall the statement of the Intermediate Value Theorem: Let $f(x)$ be a continuous function on the interval $[a, b]$. If $d \in [f(a), f(b)]$, then there is a $c \in [a, b]$ such that $f(c) = d$.

By replacing $f(x)$ by $f(x) - d$, we may assume that $d = 0$; it then suffices to obtain the following version: Let $f(x)$ be a continuous function on the interval $\in [a, b]$. If $f(a)$ and $f(b)$ have opposite signs, then there is a $c \in [a, b]$ such that $f(c) = 0$. Here is an outline of its proof: Let's assume that $f(a) < 0$, while $f(b) > 0$, the other case being handled similarly. Set $a_0 = a$ and $b_0 = b$.

Now consider the midpoint,

$$m_0 = \frac{a_0 + b_0}{2},$$

and evaluate $f(m_0)$. If $f(m_0) < 0$, set $a_1 = m_0$ and $b_1 = b_0$. If $f(m_0) > 0$, set $a_1 = a_0$ and $b_1 = m_0$. (If $f(m_0)=0$, you're already done.) What situation are we in now? $f(a_1)$ and $f(b_1)$ still have opposite signs, but the length of the interval $[a_1, b_1]$ is only half of the length of the original interval $[a_0, b_0]$. Note also that $a_0 \leq a_1$ and that $b_0 \geq b_1$.

The second step: Consider the midpoint,

$$m_1 = \frac{a_1 + b_1}{2},$$

and evaluate $f(m_1)$. If $f(m_1) < 0$, set $a_2 = m_1$ and $b_2 = b_1$. If $f(m_1) >$

0, set $a_2 = a_1$ and $b_2 = m_1$. (If $f(m_1)$=0, you're already done.) What situation are we in now? $f(a_2)$ and $f(b_2)$ still have opposite signs, but the length of the interval $[a_2, b_2]$ is only a quarter of the length of the original interval $[a_0, b_0]$. Note also that $a_0 \leq a_1 \leq a_2$ and that

$$4b_0 \geq b_1 \geq b_2.$$

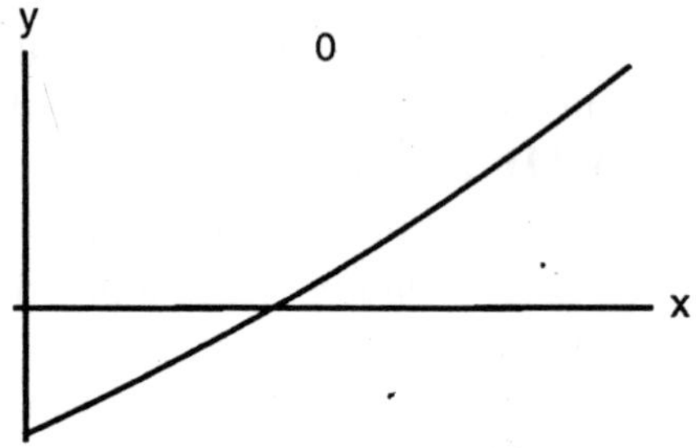

Fig. The Red Line Shows the Interval $[a_n, b_n]$.

Continuing in this fashion we construct by induction two sequences,

$$(a_n)_{n=1}^{\infty} \text{ and } (b_n)_{n=1}^{\infty}$$

with the following properties:

1. (a_n) is an increasing sequence, (b_n) is a decreasing sequence.
2. $a_n \leq b_n$ for all n.
3. $f(a_n) < 0$ for all n, $f(b_n) > 0$ for all n.
4. $b_n - a_n = 2^{-n}(b - a)$ for all n.

It follows from the first two properties that the sequences (a_n) and (b_n) converge; set

$$\lim_{x \to \infty} a_n = a, \quad \lim_{x \to \infty} b_n = b.$$

The third property and the continuity of the function $f(x)$ imply that $f(a) \leq 0$ and that $f(b) \geq 0$. The crucial observation is the fact that the fourth property implies that $a = b$. Consequently, $f(a) = f(b) = 0$, and we are done.

Example: Let's compute numerical approximations for the $\sqrt{2}$ with the help of the bisection method. We set $f(x) = x^2 - 2$. Let us start with an interval of length one: $a_0 = 1$ and $b_1 = 2$.

Note that $f(a_0) = f(1) = -1 < 0$, and $f(b_0) = f(2) = 3 > 0$. Here are the first 20 applications of the bisection algorithm:

n	a_n	a_n
0	1.	2.
1	1.	1.5
2	1.25	1.5
3	1.375	1.5
4	1.375	1.4375
5	1.40625	1.4375
6	1.40625	1.4375
7	1.4140625	1.421875
8	1.4140625	1.41796875
9	1.4140625	1.416015625
10	1.4140625	1.415039063
11	1.4140625	1.414550781
12	1.4140625	1.414306641
13	1.4148457	1.414306641
14	1.4148457	1.414245605
15	1.4148457	1.414215088
16	1.41499829	1.414215088
17	1.414207458	1.414215088
18	1.414211273	1.414215088
19	1.414213181	1.414215088
20	1.414213181	1.414214134

A comparison of the Bisection Method and the Newton-Raphson Method. The Newton-Raphson Method is often much faster than the Bisection Method. In the last example, we started with an interval of length 1.

After 10 steps, the interval $[a_{10}, b_{10}]$ has length 1/1024. Consequently every 10 steps of the Bisection Method will give us about 3 digits more accuracy - that is rather slow.

The Newton-Raphson Method can be unreliable: If the algorithm encounters a point x where $f'(x) = 0$, it crashes; if it encounters points where the derivative is very close to 0, it will become very unreliable.

The Bisection Method on the other hand will always work, once you have found starting points a and b where the function takes opposite signs.

THE DERIVATIVE

The concept of Derivative is at the core of Calculus and modern mathematics. The definition of the derivative can be approached in two different ways. One is geometrical and the other one is physical. Historically there was a fight between mathematicians which of the two illustrates the concept of the derivative best and which one is more useful. We will not dwell on this and will introduce both concepts. Our emphasis will be on the use of the derivative as a tool.

GEOMETRICAL CONCEPT OF DERIVATIVE

Consider a function $y = f(x)$ and its graph. Recall that the graph of a function is a set of points (that is $(x, f(x))$ for x's from the domain of the function f). We may draw the graph in a plane with a horizontal axis (usually called the x-axis) and a vertical axis (usually called the y-axis). Fix a point on the graph, say $(x_0, f(x_0))$. If the graph as a geometric figure is "nice" (i.e. smooth) around this point, it is natural to ask whether one can find the equation of the straight line "touching" the graph at that point.

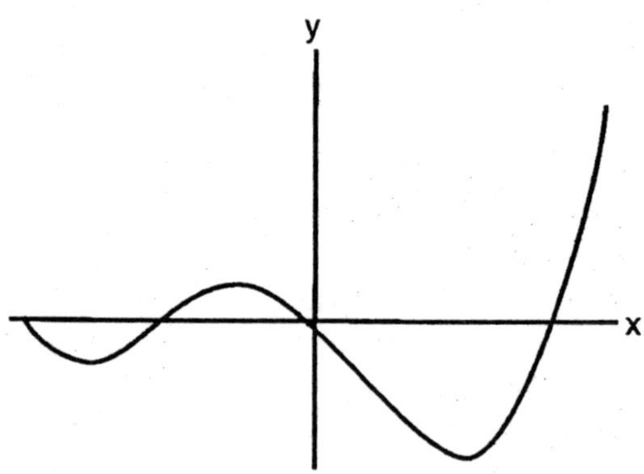

Such a straight line is called the tangent line at the point in question. The concept of tangent may be viewed in a more general framework.

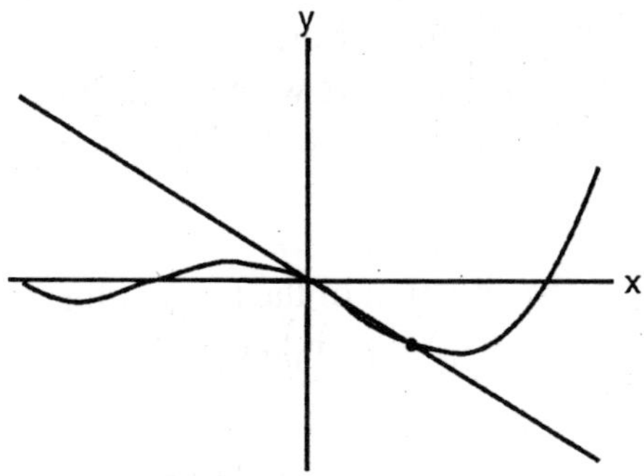

One way to find the tangent line is to consider points $(x,f(x))$ on the graph, where x is very close to x_0. Then draw the straight-line joining both points:

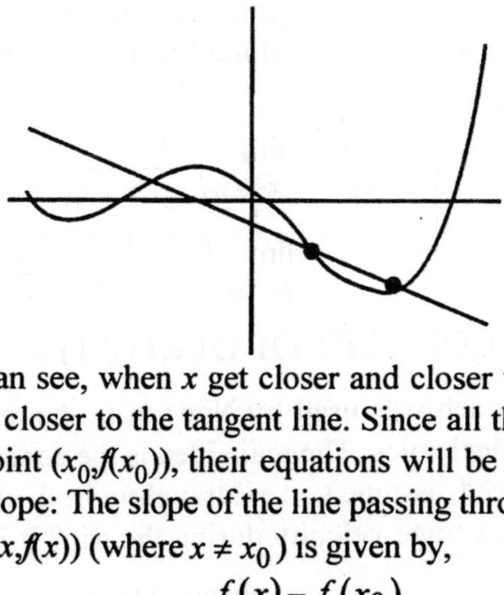

As you can see, when x get closer and closer to x_0, the lines get closer and closer to the tangent line. Since all these lines pass through the point $(x_0,f(x_0))$, their equations will be determined by finding their slope: The slope of the line passing through the points $(x_0,f(x_0))$ and $(x,f(x))$ (where $x \neq x_0$) is given by,

$$m(x) = \frac{f(x) - f(x_0)}{x - x_0}$$

The tangent itself will have a slope m, which is very close to $m(x)$ when x itself is very close to x_0. This is the concept of limit once again! In other words, we have

$$m = \lim_{x \to x_0} m(x) = \lim_{x \to x_0} \frac{f(x) - f(x_0)}{x - x_0}$$

So the equation of the tangent line is,

$$y - f(x_0) = m(x - x_0).$$

Notation. Writing "m" for the slope of the tangent line does not carry enough information; we want to keep track of the function $f(x)$ and the point x_0 in our notation. The common notation used is

$$m = f'(x_0).$$

In this case, the equation of the tangent line becomes

$$y - f(x_0) = f'(x_0)\,(x - x_0)$$

where

$$f'(x_0) = \lim_{x \to x_0} \frac{f(x) - f(x_0)}{x - x_0}$$

One last remark: Sometimes it is more convenient to compute limits when the variable approaches 0. One way to do that is to make a translation along the x-axis. Indeed, if we set $h = x - x_0$, we get

$$f'(x_0) = \lim_{x \to x_0} \frac{f(x) - f(x_0)}{x - x_0}$$

$$\lim_{h \to 0} \frac{f(x_0 + h) - f(x_0)}{h}$$

PHYSICAL CONCEPT OF DERIVATIVE

This approach was used by Newton in the development of his Classical Mechanics. The main idea is the concept of velocity and speed. Indeed, assume you are traveling from point A to point B, what is the average velocity during the trip? It is given by

$$\text{Average velocity} = \frac{\text{distance from A to B}}{\text{time to get from A to B}}$$

If we now assume that A and B are very close to each other, we get close to what is called the instantaneous velocity. Of course, if A and B are close to each other, then the time it takes to travel from A to B will also be small. Indeed, assume that at time $t=a$, we are at A. If the time elapsed to get to B is Δt, then we will be at B at time $t = a + \Delta t$.

If Δs is the distance from A to B, then the average velocity is

$$\text{Average velocity} = \frac{\Delta s}{\Delta t}.$$

The instantaneous velocity (at A) will be found when Δt get smaller and smaller.

Here we naturally run into the concept of limit. Indeed, we have,

$$\text{Instantaneous velocity (at A)} = \lim_{\Delta t \to 0} \frac{\Delta s}{\Delta t}$$

If $f(t)$ describes the position at time t, then

$$\Delta s = f(a + \Delta t) - f(a).$$

In this case, we have

$$\text{Instantaneous velocity (at A)} = \lim_{\Delta t \to 0} \frac{f(a + \Delta t) - f(a)}{\Delta t}$$

Example: Consider a parabolic motion given by the function $f(t) = t^2$.

The instantaneous velocity at $t=a$ is given by,

$$\lim_{\Delta t \to 0} \frac{f(a + \Delta t) - f(a)}{\Delta t} = \lim_{\Delta t \to 0} \frac{f(a + \Delta t)^2 - a^2}{\Delta t}$$

Since,

$$\frac{(a + \Delta t)^2 - a^2}{\Delta t} = \frac{2a\Delta t + \Delta t^2}{\Delta t} = 2a + \Delta t$$

we conclude that the instantaneous velocity at $t = a$ is $2\,a$.

This concept of velocity may be extended to find the rate of change of any variable with respect to any other variable. For example, the volume of a gas depends on the temperature of the gas.

So in this case, the variables are V (for volume) as a function of T (the temperature). In general, if we have $y = f(x)$, then the average rate of change of y with respect to x from $x = a$

to $x = a + \Delta x$, where $\Delta x \neq 0$, is,

$$\text{Average Rate} = \frac{\Delta y}{\Delta x} = \frac{f(a+\Delta x) - f(a)}{\Delta x}$$

As before, the instantaneous rate of change of y with respect to x at $x = a$, is. Instantaneous Velocity,

$$(\text{at } x = a) = \lim_{\Delta x \to 0} \frac{\Delta y}{\Delta x} = \lim_{\Delta x \to 0} \frac{f(a+\Delta x) - f(a)}{\Delta x}$$

Notation. Now we get to the hardest part. Since we can not keep on writing "Instantaneous Velocity" while doing computations, we need to come up with a suitable notation for it. If we write dx for Δx small, then we can use the notation

$$\text{Instantaneous Velocity (at } x = a) = \frac{dy}{dx}(a)$$

This is the notation introduced by Leibniz.

COMPUTATION OF DERIVATIVE

We have seen in the previous page how the derivative is defined: For a function $f(x)$, its derivative at $x = a$ is defined by,

$$f'(a) = \lim_{x \to a} \frac{f(x) - f(a)}{x - a}$$

$$\lim_{h \to o} \frac{f(a+h) - f(a)}{h}$$

Let us give some examples.

Example: Let us start with the function $f(x) = x^2$. We have

$$\frac{f(a+h) - f(a)}{h} = \frac{(a+h)^2 - a^2}{h}$$

$$= \frac{2ah + h^2}{h} = 2a + h.$$

So,

$$\lim_{h \to 0} \frac{f(a+h) - a}{h} \quad 2a$$

which means $f'(a) = 2a$.

What about the derivative of $f(x) = x^n$. Similar calculations, using the binomial expansion for $(x+y)^n$ (Pascal's Triangle), yield

$$f'(a) = na^{n-1}$$

Example: Let us consider the function,

$$f(x) = x\sin\left(\frac{1}{x}\right)$$

for $x \neq 0$, with $f(0) = 0$.

We have,

$$\frac{f(0+h)-f(0)}{h} = \frac{h\sin\left(\frac{1}{h}\right)}{h}$$

$$= \sin\left(\frac{1}{h}\right)$$

Recall that the function $\sin\left(\frac{1}{x}\right)$ has no limit when x goes to 0. So the function has no derivative and no half-derivatives as well at $x=0$.

Example: Consider the function,

$$f(x) = \sqrt{|x|}\,.$$

Then we have,

$$\frac{f(0+h)-f(0)}{h} = \frac{\sqrt{|h|}}{h}$$

$$= \frac{1}{\sqrt{|h|}}$$

Since,

$$\lim_{h\to 0} \frac{1}{\sqrt{|h|}}\text{s} = +\infty$$

then f '(0) does not exist. But observe that the graph as a geometric figure has a tangent-albeit vertical: In fact, the way the concept of the tangent line was introduced is based on the notion of slope. You already know that vertical lines do not have slopes.

So we say that the derivative does not exist whenever the tangent line is vertical.

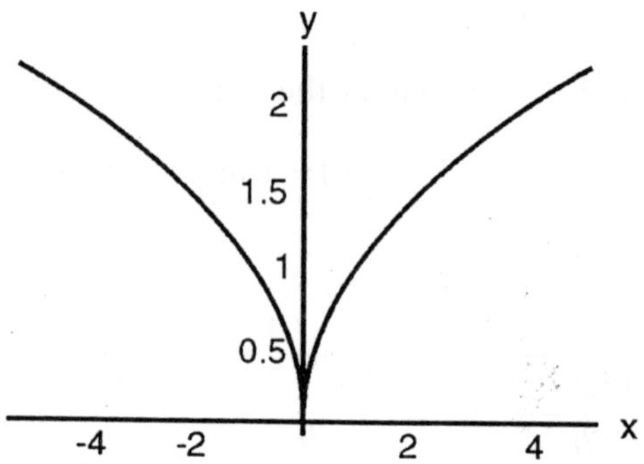

Nevertheless keep in mind that when the limit giving the derivative is $\pm\infty$ then the function has a vertical tangent line at the point.

It can be quite laborious (or impossible) to compute the derivative by hand as we have done so far. In the next pages we will show how techniques of differentiation help bypass the limit calculations and make our life much easier.

Example: Consider the function $f(x)=1/x$ for $x \neq 0$. We have

$$\frac{f(a+h)-f(a)}{h} = \frac{\frac{1}{a+h}-\frac{1}{a}}{h}$$

$$= \frac{a-(a+h)}{ha(a+h)}$$

$$= \frac{-h}{ha(a+h)}$$

Consequently,

$$\lim_{h \to 0} \frac{f(a+h)-f(a)}{h} = \lim_{h \to 0} \frac{-1}{a(a+h)} = -\frac{1}{a^2}.$$

The algebraic trick in both of the examples above has been to factor out "h" in the numerator, so that we can cancel it with the "h" in the denominator! This is what you try to do whenever you are asked to compute a derivative using the limit definition.

Example: Let us discuss the derivative of $f(x) = |x|$ at 0. We have

$$\frac{f(0+h)-f(a)}{h} = \frac{|h|-0}{h}$$
$$= \frac{|h|}{h}$$

But

$$\lim_{h\to 0+} \frac{|h|}{h} = 1 \text{ and } \lim_{h\to 0-} \frac{|h|}{h} = -1$$

which implies that $f'(0)$ does not exist.

Remark. This example is interesting. Even though the derivative at the point does not exist, the right and the left limit of the ratio do exist. In fact, if we use the slope-interpretation of the derivative we see that this means that the graph has two lines close to it at the point under consideration. They could be seen as "half-tangents". See Picture.

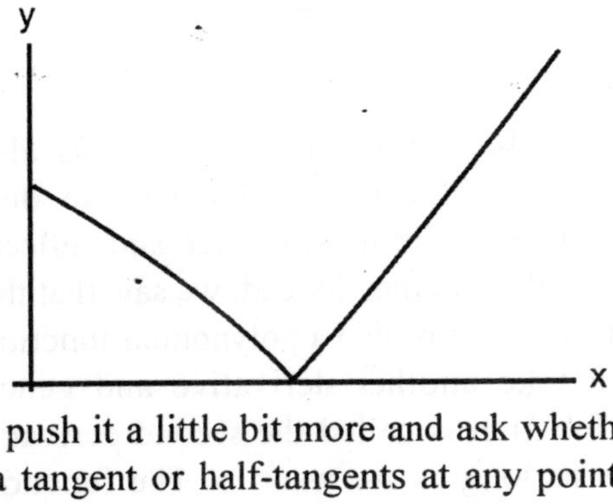

So let's push it a little bit more and ask whether a function always has a tangent or half-tangents at any point. That is not the case either.

Techniques of Differentiation

Maybe the easiest and most useful formulas are the ones that say that the derivative is linear:

$$(f+g)'(a) = f'(a)+g'(a)$$
$$(cf)'(a) = cf'(a)$$

Combined with the formula $(x^n)' = n\,x^{n-1}$, we see that every polynomial function has a derivative at any point.

Example: For $P(x) = 1-2x + 3x^4 - 5\,x^6$, we have

$$P'(x) = -2+12x^3-30x^5$$

The next two formulas are the most powerful ones. They deal with the derivative of a product and a quotient. They are commonly called the product rule and the quotient rule. We have

$$(f\,g)'(a) = f'(a)g(a) + f(a)g'(a)$$

$$\left(\frac{f}{a}\right)'(a) = \frac{f'(a)g(a) - f(a)g'(a)}{g^2(a)}$$

In particular, we have

$$\left(\frac{1}{f}\right)'(a) = -\frac{f'(a)}{f^2(a)}$$

So, we have

$$\left(\frac{1}{x^n}\right) = -\frac{nx^{n-1}}{x^{2n}}$$
$$= -nx^{-n-1}$$

which means that the formula $(x^r)' = r\,x^{r-1}$ is also valid for negative exponents. Before we discuss the derivative of trigonometric functions, let us stop here and reflect a little bit more on polynomial functions. Indeed, we saw that the derivative of a polynomial function is also a polynomial function.

So we can take another derivative and generate a new function. This function is called the second derivative. We can keep doing this as long as we want to. The functions obtained are called higher derivatives. The common notations used for them are

$$f'(x),\quad f''(x),\quad f'''(x),\quad f^{(4)}(x),\quad \ldots,\quad f^{(n)}(x)$$

$$y',\quad y'',\quad y''',\quad y^{(4)},\quad \ldots,\quad y^{(n)}$$

$$\frac{dy}{dx},\quad \frac{d^2y}{dx^2},\quad \frac{d^3y}{dx^3},\quad \frac{d^4y}{dx^2}\quad \ldots,\quad \frac{d^ny}{dx^n}$$

The Derivatives of Trigonometric Functions

$$\sin' x = \cos x \qquad \cos' x = -\sin x$$

$$\tan' x = \sec^2 x \qquad \cot' x = -\csc^2 x$$

$$\sec' x = \sec x \tan x \quad \csc' x = -\csc x \cot x$$

Trigonometric functions are useful in our practical lives in diverse areas such as astronomy, physics, surveying, carpentry etc.

How can we find the derivatives of the trigonometric functions?

Our starting point is the following limit:

$$\lim_{x \to 0} \frac{\sin(x)}{x} = 1$$

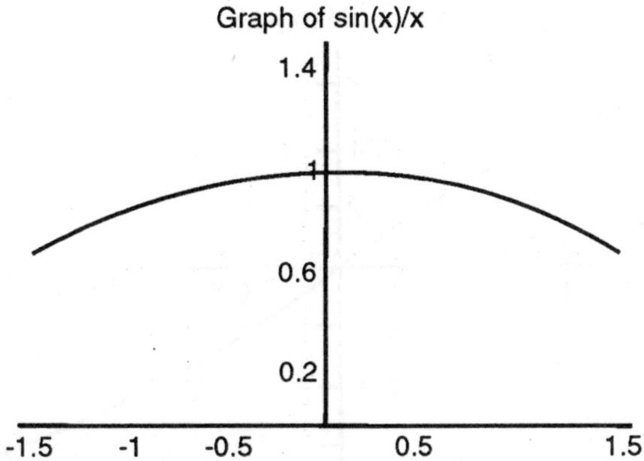

Using the derivative language, this limit means that $\sin'(0) = 1$. This limit may also be used to give a related one which is of equal importance:

$$\lim_{x \to 0} \frac{\cos(x) - 1}{x} = 1$$

To see why, it is enough to rewrite the expression involving the cosine as,

$$\frac{\cos(x) - 1}{x} = \frac{(\cos(x) - 1)(\cos(x) + 1)}{x(\cos(x) + 1)}$$

$$= \frac{\left(\cos^2(x)-1\right)}{x\left(\cos(x)+1\right)}$$

But,

$$\cos^2(x)-1=-\sin^2(x),$$

so we have

$$\lim_{x\to 0}\frac{\cos(x)-1}{x}=\lim_{x\to 0}\frac{-\sin^2(x)}{x\left(\cos(x)+1\right)}$$

$$=\lim_{x\to 0} x\frac{-\sin^2(x)}{x^2\left(\cos(x)+1\right)}=0$$

This limit equals cos'(0) and thus cos'(0) = 0.

Graph of (cos(x)-1/x

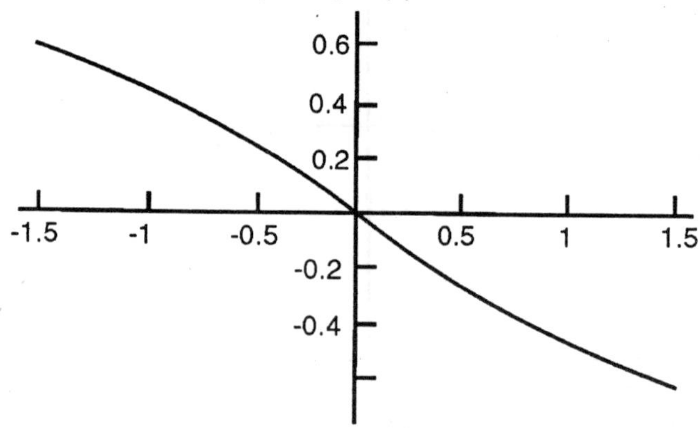

In fact, we may use these limits to find the derivative sin (x) of and sin (x) at any point $x = a$. Indeed, using the addition formula for the sine function, we have

$$\sin(a+h) = \sin(a)\cos(h)+\sin(h)\cos(a)$$

So,

$$\frac{\sin(a+h)}{h} = \frac{\sin(a)\cos(h)-1}{h}+\cos(a)\frac{\sin(h)}{h}$$

which implies,

$$\lim_{h \to 0} \frac{\sin(a+h) - \sin(a)}{h} = \cos(a)$$

So we have proved that sin'(a) exists and sin'(a) = cos(a). Similarly, we obtain that cos' (a) exists and that cos'(a) = –sin(a). Since tan(x), cot(x), sec (x), and csc(x)are all quotients of the functions and, we can compute their derivatives with the help of the quotient rule:

$$\frac{d}{dx}(\tan(x)) = \sec^2(x) = 1 + \tan^2(x)$$

$$\frac{d}{dx}(\sec(x)) = \sec(x) = 1 + \tan(x)$$

$$\frac{d}{dx}(\tan(x)) = -\csc^2(x) = -1 - \cot^2(x)$$

$$\frac{d}{dx}(\csc(x)) = -\csc(x)\cot(x)$$

It is quite interesting to see the close relationship between tan (x) and (and also between cot (x) and csc (x)).

From the above results we get

$$\sin''(x) = -\sin(x) \text{and} \cos'' = -\cos(x).$$

These two results are very useful in solving some differential equations.

Example: Let $f(x) = \sin(2x)$. Using the double angle formula for the sine function, we can rewrite

$$\sin(2x) = 2\sin(x)\cos(x)$$

So using the product rule, we get

$$\frac{d}{dx}(\sin(2x)) = 2(\cos(x)\cos(x) - \sin(x)\sin(x))$$

$$= 2(\cos^2(x) - \sin^2(x))$$

which implies, using trigonometric identities,

$$\frac{d}{dx}(\sin(2x)) = 2\cos(2x)$$

In fact next we will discuss a formula which gives the above conclusion in an easier way.

Implicit Differentiation

In many examples, especially the ones derived from differential equations, the variables involved are not linked to each other in an explicit way. Most of the time, they are linked through an implicit formula, like $F(x,y)=0$. Once x is fixed, we may find y through numerical computations. (By some fancy theorems, we may formally show that y may indeed be seen as a function of x over a certain interval). The question becomes what is the derivative $\frac{dy}{dx}$, at least at a certain a point? The method of implicit differentiation answers this concern. Let us illustrate this through the following example.

Example: Find the equation of the tangent line to the ellipse

$$25\,x^2 + y^2 = 109$$

at the point (2,3). One way is to find y as a function of x from the above equation, then differentiate to find the slope of the tangent line.

We will leave it to the reader to do the details of the calculations. Here, we will use a different method. In the above equation, consider y as a function of x:

$$25\,x^2 + y(x)^2 = 109,$$

and use the techniques of differentiation, to get

$$\begin{aligned}\frac{d}{dx}\left[25x^2 + y(x)^2\right] &= 50x + 2y\frac{dy}{dx} \\ &= 0\end{aligned}$$

From this, we obtain

$$\frac{dy}{dx} = -25\frac{x}{y}$$

which implies that $\frac{dy}{dx} = -\frac{50}{3}$ at the point (2,3). So the equation of the tangent line is given by,

$$y-3 = -\frac{50}{3}(x-2),$$

or equivalently,

$$3y + 50\,x = 109.$$

You may wonder why bother if this is just a different way of finding the derivative? Consider the following example! It can be very hard or in fact impossible to solve explicitly for y as a function of x.

Example: Find y' if

$$y^4 + 4y - 3x^3 \sin(y) = 2x+1$$

This is a wonderful example of an implicit relation between $x$and y. So how do we find y'?

Let us differentiate the above equation with respect to x where y is considered to be a function of x. We get

$$4y^3 y' + 4y' - 9x^2 \sin(y) - 3x^3 \cos(y)\,y' = 2$$

Easy algebraic manipulations give

$$y' = \frac{2+9x^2\sin(y)}{4y^3+4-3x^3\cos(y)}$$

We can also find higher derivatives of y such as y'' in this manner. We only have to differentiate the above result. Of course the calculations get little more messy.

The Chain Rule

$$(f\,og(x))' = g'(x)\,f'(g(x))$$

or
$$\frac{dy}{dx} = \frac{dy}{du}\frac{du}{dx}$$

As a motivation for the chain rule, consider the function

$$f(x) = (1+x^2)^{10}.$$

Since $f(x)$ is a polynomial function, we know from previous pages that $f'(x)$ exists.

Naturally one may ask for an explicit formula for it. One tedious way to do this is to develop $(1+x^2)^{10}$ using the Binomial Formula and then take the derivative. Of course, it is possible to do this, but it won't be much fun. But what if we have to deal with $(1+x^2)^{100}$! Then I hope you agree that the Binomial Formula is not the way to go anymore. So what do we do? The answer is given by the Chain Rule.

Before we discuss the Chain Rule formula, let us give another example.

Example: Let us find the derivative of $f(x) = \sin(2x)$. One way to do that is through some trigonometric identities. Indeed, we have

$$\sin(2x) = 2\sin(x)\cos(x)$$

So we will use the product formula to get

$$(\sin(2x))' = 2(\sin'(x)\cos(x) + \sin'(x)\cos'(x))$$

which implies,

$$(\sin(2x))' = 2(\cos^2(x) - \sin^2(x))$$

Using the trigonometric formula

$$\cos(2x) = \cos^2(x) - \sin^2(x), \text{ we get}$$

$$(\sin(2x))' = 2\cos(2x)$$

Once this is done, you may ask about the derivative of sin $(5x)$? The answer can be found using similar trigonometric identities, but the calculations are not as easy as before. Again we will see how the Chain Rule formula will answer this question in an elegant way.

In both examples, the function $f(x)$ may be viewed as:

$$f(x) = h(g(x))$$

where $g(x) = 1+x^2$ and $h(x) = x^{10}$ in the first example, and $h(x) = \sin(x)$and $g(x) = 2x$ in the second. We say that $f(x)$ is the composition of the functions $g(x)$ and $h(x)$ and write

$$f(x) = hog(x)$$

The derivative of the composition is given by the formula

$$(fog(x))' = g'(x)f'(g(x))$$

Another way to write this formula is,

$$\frac{dy}{dx} = \frac{dy}{du}\frac{du}{dx}$$

where,

$$y = f \circ g(x) = f(u)$$

and $u = g(x)$. This second formulation (due to Leibniz) is easier to remember and is the formulation used almost exclusively by physicists.

Example: Let us find the derivative of

$$f(x) = \left(1+x^2\right)^{100}$$

We have,

$$f(x) = h \text{o} g(x),$$

where,

$$g(x) = 1+x^2 \text{ and } h(x) = x^{100}.$$

Then the Chain rule implies that $f'(x)$ exists, which we knew since it is a polynomial function, and,

$$\begin{aligned} f'x &= 2x \cdot \left[100\left(1+x^2\right)^{99}\right] \\ &= 200x\left(1+x^2\right)^{99} \end{aligned}$$

Example: Let us find the derivative of

$$f(x) = \sin(5x)$$

We have $f(x) = h \text{o} g(x)$, where

$$g(x) = 5x \text{ and}$$
$$h(x) = \sin(5x).$$

Then the Chain rule implies that $f'(x)$ exists and

$$\begin{aligned} f'(x) &= 5 \cdot [\cos(5x)] \\ &= 5\cos(5x) \end{aligned}$$

In fact, this is a particular case of the following formula

$$[f(ax+b)]' = af'(ax+b)$$

The following formulas come in handy in many areas of techniques of integration.

$$\frac{d}{dx}\left(u^n\right) = nu^{n-1}\frac{du}{dx}$$

$$\frac{d}{dx}(\cos(u)) = -\sin(u)\frac{du}{dx}$$

$$\frac{d}{dx}(\sin(u)) = \cos(u)\frac{du}{dx}$$

$$\frac{d}{dx}(\mathrm{In}(u)) = \frac{1}{u}\frac{du}{dx}$$

$$\frac{d}{dx}\left(d^u\right) = e^u\frac{du}{dx}$$

More formulas for derivatives can be found in our section of tables.

Linear Approximations

This approximation is crucial to many known numerical techniques such as Euler's Method to approximate solutions to ordinary differential equations. The idea to use linear approximations rests in the closeness of the tangent line to the graph of the function around a point. Let x_0 be in the domain of the function $f(x)$.

The equation of the tangent line to the graph of $f(x)$ at the point (x_0,y_0), where $y_0 = f(x_0)$, is,

$$y - y_0 = f'(x_0)(x - x_0)$$

If x_1 is close to x_0, we will write $x_1 = x_0 + \Delta x$, and we will approximate $f(x_0 + \Delta x)$ by the point (x_1,y_1) on the tangent line given by

$$y_1 = y_0 + \Delta x f'(x_0)$$

If we write $\Delta y = y_1 - y_0$, we have

$$\Delta y = \Delta x f'(x_0)$$

In fact, one way to remember this formula is to write $f'(x)$ as $\frac{dy}{dx}$ and then replace d by Δ. Recall that, when x is close to x_0, we have,

$$f(x) \approx f(x_0) + f'(x_0)(x - x_0)$$

Example: Estimate $\sqrt{9.2}$.

Let $f(x) = \sqrt{3x}$. We have $f(6) = \sqrt{9} = 3$. Using the above approximation, we get

$$f(6.2) \approx f(6) + f'(6)(6.2 - 6)$$

We have

$$f'(x) = \frac{1}{2}\frac{1}{\sqrt{x+3}}$$

So $f'(6) = \frac{1}{6}$. Hence

$$f(6.2) \approx f(6) + f'(6)(6.2 - 6) = 3.033$$

or,

$$\sqrt{9.2} \approx 3.033.$$

Check with your calculator and you'll see that this is a pretty good approximation for $\sqrt{9.2}$.

Remark: For a function $f(x)$, we define the differential df of $f(x)$ by

$$df = f'(x)dx.$$

Example: Consider the function $y = f(x) = 5x^2$. Let Δx be an increment of x. Then, if Δy is the resulting increment of y, we have

$$\begin{aligned} \Delta y &= f(x + \Delta x) - f(x) \\ &= 5(x + \Delta x)^2 - 5x^2 \\ &= 10x(\Delta x) + 5(\Delta x)^2 \end{aligned}$$

On the other hand, we obtain for the differential dy:

$$dy = f'(x)dx$$

$$= 10xdx$$

In this example we are lucky in that we are able to compute Δy exactly, but in general this might be impossible. The error in the approximation, the difference between dy(replacing dx by Δx) and Δy, is $5(\Delta x)^2$, which is small compared to Δx.

5

Limit Theorem

LIMIT

In mathematics, the concept of a "limit" is used to describe the behavior of a function as its argument either "gets close" to some point, or as it becomes arbitrarily large; or the behavior of a sequence's elements as their index increases indefinitely. Limits are used in calculus and other branches of mathematical analysis to define derivatives and continuity. The concept of the "limit of a function" is further generalized to the concept of topological net, while the limit of a sequence is closely related to limit and direct limit in category theory.

LIMIT OF A SEQUENCE

Th concept on which the whole of analysis ultimately rests is that of the limit of a sequence. We shall first make the position clear by considering several examples.

$a_n = 1/n$: We consider the sequence

$$a_1 = 1, a_2 = \frac{1}{2}, a_3 = \frac{1}{3}, \ldots, a_n = \frac{1}{n}, \ldots$$

No number of this sequence is zero; but we see that the larger the number n, the closer to zero is the number a_n. Hence, if we mark off around the point 0 an interval as small as we please, then starting with a definite value of the subscript all the numbers a_n will fall into this interval. We express this state of affairs by saying that, as n increases, the numbers a_n tend to 0, or that they possess the limit 0, or that the sequence a_1, a_2, a_3, converges to 0.

If numbers are represented as points on a line, this means that the points $1/n$ crowd closer and closer to the point 0 as n increases.

The situation is similar in the case of the sequence

$$a_1 = 1, a_2 = -\frac{1}{2}, a_3 = \frac{1}{3}, a_4 = -\frac{1}{4}, \ldots, a_n = \frac{(-1)^{n-1}}{n}, \ldots$$

Here too, the numbers a_n tend to zero as n increases; the only difference is that the numbers a_n are sometimes larger and sometimes smaller than the limit 0; as we say, they oscillate about the limit.

The convergence of the sequence to 0 is usually expressed symbolically by the equation

$$\lim_{x \to \infty} a_n = 0,$$

or occasionally by the abbreviation

$$a_n \to 0.$$

a_{2m} = 1/m; a_{2m-1} = 1/2m: In the preceding examples, the absolute value of the difference between a_n and the limit steadily becomes smaller as n increases. This is not necessarily always the case, as is shown by the sequence

$$a_1 = \frac{1}{2}, a_2 = 1, a_3 = \frac{1}{4}, a_4 = \frac{1}{2}, a_5 = \frac{1}{6}, a_6 = \frac{1}{3}, \ldots;$$

that is, in general, for even values $n = 2m$, $a_n = a_{2m}$, $= 1/m$, for odd values $n = 2m - 1$, $a_n = a_{2m-1} = 1$. This sequence also has a limit, namely zero, since every interval about the origin, no matter how small, will contain all the numbers a_n from a certain value of n onwards; but it is not true that every number lies nearer to the limit zero than the preceding one.

$a_n = n/(n + 1)$: We consider the sequence

$$a_1 = \frac{1}{2}, a_2 = \frac{2}{3}, \ldots a_n = \frac{n}{n+1}, \ldots,$$

where the integral subscript n takes all the values 1, 2, 3, ... If we write $a_n = 1 - 1/(n + 1)$, we see at once that, as n increases, the numbers a_n will approach closer and closer to the number 1, in the sense that, if we mark off any interval about the point 1, all the numbers a_n following a certain a_N must fall into that interval. We write

$$\lim_{x \to \infty} a_n = 1.$$

The sequence

$$a_n = \frac{n^2 - 1}{n^2 + n + 1}$$

behaves in a similar manner. This sequence also tends to a limit as n increases to the limit 1, in fact, in symbols, $\lim_{x \to \infty} a_n = 1$. We see this most readily, if we write

$$a_n = 1 - \frac{n+2}{n^2 + n + 1} = 1 - r_n;$$

now we have only to show that the numbers r_n tend to 0 as n increases. For all values of n greater than 2, we have $n + 3 < 2n$ and $n + n + 1 > n$. Hence we have for the remainder r_n

$$0 < r_n < \frac{2n}{n^2} = \frac{2}{n} (n > 2),$$

which shows at once that r_n tends to 0 as n increases. Our discussion yields at the same time an estimate of the amount by which the number a_n (for $n > 2$) can at most differ from the limit 1; this difference certainly cannot exceed 2/n.

This example illustrates the fact, which we should naturally expect, that for large values of n the terms with the highest indices in the numerator and denominator of the fraction for a_n predominate and that they determine the limit.

$a_n = \sqrt[n]{p}$ Let p be any fixed positive number. We consider the sequence $a_1, a_2, a_3, \ldots, a_n,$, where

$$a_n = \sqrt[n]{p}$$

We assert that

$$\lim_{n \to \infty} a_n = \lim_{n \to \infty} \sqrt[n]{p} = 1$$

We can prove this very easily by using a lemma which we shall find also useful for other purposes.

If $1 + h$ is a positive number (that is, if $h > -1$) and n is an integer greater than 1, then

$$(1 + h)n > 1 + nh.......$$

Assume that Inequality already has been proved for a certain $m > 1$; multiply both sides by $(1+h)$ and obtain

$$(1 + h)^{m+1} > (1 + mh)(1 + h) = 1 + (m + 1)h + mh^2$$

If we omit on the right hand side the positive term mh, the inequality remains valid. We thus obtain

$$(1 + h)^{m+1} > 1+(m + 1)h$$

However, this is our inequality for the index $m + 1$. Hence, if the inequality holds for the index m, is also holds for $m + 1$. Since it holds for $m = 2$, it also holds for $m = 3$, whence for $m = 4$, and so on, whence it holds for every index. This is a simple example of a proof by mathematical induction, a type of proof which is often useful. Returning to our sequence, we distinguish between the case $p > 1$ and the case $p < 1$ (if $p = 1$, then $\sqrt[n]{p}$ is also equal to 1 for every n and our statement becomes trivial). If $p > 1$, then $\sqrt[n]{p}$ will also be greater than 1; let $\sqrt[n]{p} = 1 + h_n$, where h_n is a positive quantity depending on n and we find by Inequality

$$p = (1+h_n)n > 1+nh_n$$

whence follows at once that

$$0 < h_n < \frac{p-1}{n}$$

Thus, as n increases, the number h_n must tend to 0, which proves that the numbers a_n converge to the limit 1, as stated. At the same time, we have a means for estimating how close any a_n is to the limit 1; the difference between a_n and 1 is certainly not greater than $(p — 1)/n$. If $p < 1$, then $\sqrt[n]{p}$ will likewise be less than 1 and therefore may be taken equal to $1/(1 + h_n)$, where h_n is a positive number. It follows from this, using Inequality, that

$$p = \frac{1}{(1+h_n)^n} < \frac{1}{1+nh_n}$$

(By decreasing the denominator, we increase the fraction. It follows that

$$1+nh_n<\frac{1}{p},$$

whence $$h_n<1\frac{1/p-1}{n}$$

This shows that h_n tends to 0 as n increases. As the reciprocal of a quantity tending to 1, $\sqrt[n]{p}$ it self tends to 1.

$$\alpha_n=\alpha^n:$$

We consider the sequence $a_n = \alpha^n$, where α is fixed and n runs through the sequence of positive integers. First, let α be a positive number less than 1. We may then put $\alpha = 1/(1 + h)$, where h is positive and Inequality yields

$$\alpha_n=\frac{1}{(1+h)^n}<\frac{1}{1+nh}<\frac{1}{nh}$$

Since the number h and, consequently, $1/h$ depends only on n and does not change as n increases, we see that as n increases α^n tends to 0:

$$\lim_{n\to\infty}\alpha^n=0\ (0<a<1)$$

The same relationship holds when α is zero or negative, but greater than -1. This is immediately obvious, since in any case

$$\lim_{n\to\infty}|\alpha|^n=0$$

If $\alpha = 1$, then α^n will obviously be always equal to 1 and we shall have to regard the number 1 as the limit of α^n.

If $\alpha > 1$, we set $\alpha = 1 + h$, where h is positive and see at once from our inequality that, as n increases, α^n does not tend to any definite limit, but increases beyond all bounds. We express this state of affairs by saying that α^n tends to infinity as n increases, or that α^n becomes infinite; in symbols

$$\lim_{n\to\infty}\alpha^n=\infty\qquad(\alpha>1)$$

Nevertheless, as we must explicitly emphasize, the symbol ∞ does not denote a number with which we can calculate as with any other number; equations or statements which express that a quantity is or becomes infinite never have the same sense as an equation between definite quantities.

In spite of this, such modes of expression and the use of the symbol are extremely convenient.

If $\alpha = -1$, the values of α^n will not tend to any limit, but, as n runs through the sequence of positive integers, it will take the values $+1$ and -1 alternately. Similarly, if $\alpha < -1$, the value of α^n will increase numerically beyond all bounds, but its sign will be in sequence positive and. negative.

LIMITS OF A^N

If we consider the curves $y = x^n$ and $y = x^{1/n} = \sqrt[n]{x}$ and restrict ourselves, for the sake of convenience, to non-negative values of x, the preceding limits are illustrated by Figures respectively. In the case of the curves $y = x^n$, we see that in the interval 0 to 1 they approach closer and closer to the x-axis as n increases, while outside that interval they climb more and more steeply and draw in closer and closer to a line parallel to the y-axis. All the curves pass through the point with co-ordinates $x=1, y=1$ and through the origin.

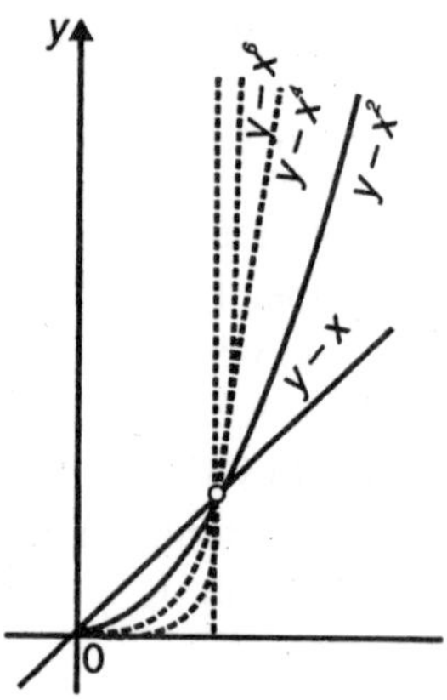

Fig. x^n as n increases

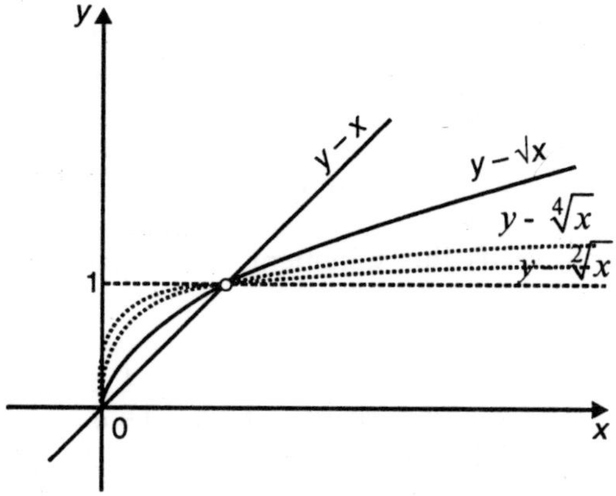

Fig. $x^{1\backslash n}$ as n increases

In the case of the functions $y = x^{1/n} = +\sqrt[n]{x}$, the curves approach closer and closer to the line parallel to the x-axis and at a distance 1 above it. On the other hand, all the curves must pass through the origin. Hence, in the limit, the curves approach the broken line consisting of the part of the y-axis between the points $y = 0$ and $y = 1$ and of the parallel to the x-axis $y = 1$. Moreover, it is clear that the two figures are closely related, as one would expect from the fact that the functions $y = \sqrt[n]{x}$ are actually the inverse functions of the n-th powers, from which we infer that each figure is transformed into the other on reflection in the line $y = x$.

GEOMETRIC SERIES

An example of a limit which is more or less familiar from elementary mathematics is the geometric series

$$1 + q + q^2 + ... + q^{n-1} = S_n;$$

the number q is called the common ratio of the series. The value of this sum may, as is well known, be expressed in. the form

$$S_n = \frac{1-q^n}{1-q},$$

provided that $q > 1$; we can derive this expression by multiplying the sum S_n by q and subtracting the equation thus obtained from the original equation, or we may verify the formula by division.

There arises now the question what happens to the sum S_n when n. increases indefinitely? The answer is: The sum S_n has a definite limit S if q lies between -1 and +1, these end values being excluded, and it is then true that

$$S = \lim_{n\to\infty} S_n = \frac{1}{1-q}.$$

In order to verify this statement, we write the numbers S_n in the form

$$S_n = \frac{1-q^n}{1-q} = \frac{1}{1-q} - \frac{q^n}{1-q}.$$

We have already shown that, provided $|q| < 1$, the quantity q^n and with it $q^n/(1-q)$ tends to 0 as n increases; hence, with the above assumption, the number S_n tends, as was stated, to the limit $1/(1-q)$ as n increases.

The passage to the limit

$$\lim_{n\to\infty}\left(1+q+q^2+...+q^{n-1}\right) = \frac{1}{1-q}$$

is usually expressed by saying that when $|q| < 1$, the geometric series can be extended to infinity and that the sum of the infinite geometric series is the expression $1/(1-q)$.

The sums S_n of the finite geometric series are also called the partial sums of the infinite geometric series $1 + q + q + ...$ (We must draw a sharp distinction between the sequence of numbers $S_1, S_2, ...,$ and the geometric series.)

The fact that the partial sums S_n of a geometric series tend to the limit $S = 1/(1-q)$ as n increases may also be expressed by saying that the infinite geometric series $1 + q + q + ...$ converges to the sum $S = 1/(1-q)$ when $|q| < 1$.

$a_n = \sqrt[n]{n}$:

We shall show that the sequence of numbers

$$a_1 = 1, a_2 = \sqrt{2}, a_3 = \sqrt[3]{3}, ..., a_n = \sqrt[n]{n}, ...$$

tends to 1 as n increases, i.e., that

$$\lim_{n\to\infty} \sqrt[n]{n} = 1.$$

Here we make use of a slight artifice. Instead of the sequence $a_n = \sqrt[n]{n}$, we first consider the sequence,

$$b_n = \sqrt{a_n} = \sqrt{\sqrt[n]{n}} = \sqrt[n]{\sqrt{n}}.$$

When $n > 1$, the term b_n is also greater than 1. We can therefore set $b_n = 1 + h_n$, where h_n is positive and depends on n. We therefore have

$$\sqrt{n} = (b_n)^n = (1 + h_n)^n \geqq 1 + nh_n,$$

so that

$$h_n \leqq \frac{\sqrt{n} - 1}{n} \leqq \frac{\sqrt{n}}{n} = \frac{1}{\sqrt{n}}.$$

We now have

$$1 \leqq a_n = b_n^2 = 1 + 2h_n + h_n^2 \leqq 1 + \frac{2}{\sqrt{n}} + \frac{1}{n}.$$

Obviously, the right hand side of this inequality tends to 1, and so does a_n.

$$a_n = \sqrt{n+1} - \sqrt{n}$$

We assert that

$$\lim_{n \to \infty} \left(\sqrt{n} + 1 - \sqrt{n} \right) = 0.$$

In order to prove this formula, we need only write this expression in the form

$$\sqrt{n+1} - \sqrt{n} = \frac{\left(\sqrt{n+1} - \sqrt{n}\right)\left(\sqrt{n+1} + \sqrt{n}\right)}{\sqrt{n+1} + \sqrt{n}} = \frac{1}{\sqrt{n+1} + \sqrt{n}};$$

we see at once that this expression tends to 0 as n increases.

$a_n = \dfrac{n}{\alpha^n}$: Let α be a number greater than 1.

We assert that as n increases the sequence of the numbers $a_n = n/a_n$ tends to the limit 0.

As in the case of $\sqrt[n]{n}$ above, we consider the sequence

$$\sqrt{a_n} = \frac{\sqrt{n}}{\left(\sqrt{\alpha}\right)^n}$$

We set $\sqrt{\alpha} = 1 + h$ Here $h > 0$, since α and hence α is greater than 1. By Inequality, we have

$$\sqrt{\alpha} = (1+h)^n > 1 + nh,$$

so that

Hence $$\sqrt{\alpha_n} = \frac{\sqrt{n}}{(1+h)^n} \leq \frac{\sqrt{n}}{1+nh} \leq \frac{\sqrt{n}}{nh} = \frac{1}{h\sqrt{n}}$$

$$a_n \leq \frac{1}{nh^2}$$

LIMIT OF FUNCTIONS

We first met a sequence as a particularly easy sort of function, defined on $\mathbb{N}$, rather than $\mathbb{R}$. We now move to the more general study of functions. However, our earlier work wasn't a diversion; we shall see that sequences prove a useful tool both to investigate functions, and to give an idea of appropriate methods. Our main reason for being interested in studying functions is as a model of some behaviour in the real world.

Typically a function describes the behaviour of an object over time, or space. In selecting a suitable class of functions to study, we need to balance generality, which has chaotic behaviour, with good behaviour which occurs rarely. If a function has lots of good properties, because there are strong restrictions on it, then it can often be quite hard to show that a given example of such a function has the required properties.

Conversely, if it is easy to show that the function belongs to a particular class, it may be because the properties of that class are so weak that belonging may have essentially no useful consequences. It turns out that there are a number of "good" classes of functions which are worth studying.

In this chapter and the next ones, we study functions which have steadily more and more restrictions on them. Which means the behaviour steadily improves; and at the same time, the number of examples steadily decreases. A perfectly general function has essentially nothing useful that can be said about it; so we start by studying continuous functions, the first class that gives us much theory. In order to discuss functions sensibly, we often insist that we can "get a good look" at the behaviour of the function at a given point, so typically we restrict the domain of the function to be well behaved.

Definition: A subset U of $\mathbb{R}$ is open if given $a \in U$, there is some $\delta > 0$ such that $(a - \delta, a + \delta) \subseteq U$.

In fact this is the same as saying that given $a \in U$, there is some open interval containing a which lies in U. In other words, a set is open if it contains a neighbourhood of each of its points. We saw in that an open interval is an open set. This definition has the effect that if a function is defined on an open set we can look at its behaviour near the point a of interest, from both sides.

Functions And Mapping

Definition: Let A and B be any two non-empty sets. Let A = $\{p, q, r\}$ and $B = \{a, b, c, d\}$. Suppose by some rule or other, we assign to each element of A a 'unique' element of B. Let p be associated to a, q be associated to b, r be associated to c etc. The set $\{(p, a), (q, b), (r, c)\}$ is called a function from set A to set B. If we denote this set by 'f' then we write f: $A \to B$ which is read as "f" is a function of A to B or 'f' is a mapping from A to B.

Definition: Let A and B be two non-empty sets. Suppose there exists a rule of correspondence, denoted by f which associates to each member of A, a unique member of B. Then 'f' is called a function or a mapping from set A to set B.

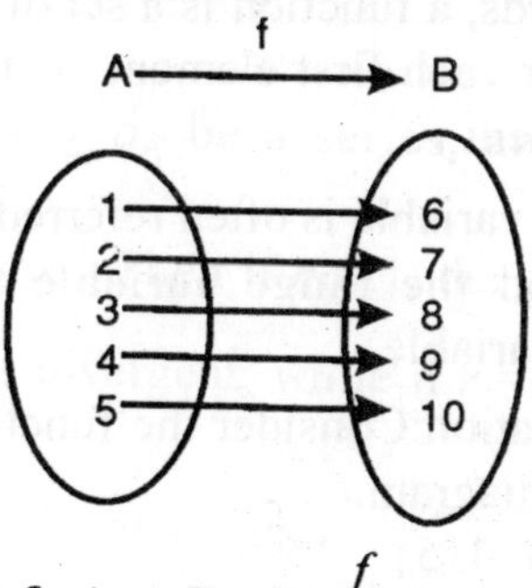

It is denoted by f: $A \to$ B or $A \xrightarrow{f}$ B

Note that:

1. The set A is called "domain" of 'f'
2. The set B is called "codomain" of 'f'
3. The element $y \in B$ which the function 'f' associates to the element $x \in A$ is denoted by $f(x)$ i.e.

   $y = f(x)$ and y is called the 'f' image of x or value of 'f' at x. The element x is also called as pre-image of y. Every element of A has a unique image but each element of B need not have as the image of an element in A. There can be more than one element of A which has the same image in B.
4. The set of these images in B is called the range of 'f' i.e. the range of a function as the images of its domain. We denote the range of $f: A \to B$ by f (A). Thus $f(A) = \{f(x): x \in A\} \Rightarrow f(A) \subseteq B$.
5. If A and B are any two non-empty sets, then a function f from A to B is a subset f of $A \times B$ satisfying the following conditions:

   a $\forall x \in A, (x, y) \in f$ for some $y \in B$

   b. $(x, y) \in f$ and $(x, y_1) \in f \Rightarrow y = y1$

   The first condition ensures that there is a rule which assigns to each element $x \in A$ some element $y \in B$. Thus each element in A will have an image. The second condition guarantees that the image is unique.

   In other words, a function is a set of ordered pair (x, y), such that for each first element 'x' there corresponds a unique element y.
6. The domain variable is often referred as the independent variable and the range variable is referred as the dependent variable.

Formula or Equation:Consider the function 'f' exhibited by the adjoining arrow diagram.

Let A = { 1, 2, 3, 4, 5}

B = { 5, 7, 9, 11, 13, 15}.

Note that, if we take any element $x \in A$ then the element of the co-domain set B related to x is obtained by adding 3 to twice of x. Applying this rule we get,

$$f(1) = 2\,(1) + 3 = 5, f(2) = 2\,(2) + 3 = 7,$$
$$f(3) = 2\,(3) + 3 = 9 \text{ etc.}$$

In general we can write $f(x) = 2x + 3, x \in A$

This is the formula which exhibits the function 'f'. If we denote the value of f at x by, $y \in B$, we get $y = 2x + 3\ x \in A$. This becomes an equation which exhibits the function f.

Remarks:

1. If a function is exhibited by a formula, then using this formula we can find the range of the function.
2. Sometimes, the domain is not known or is not given. In such a case, the domain is taken as that set of elements at which the values of the function can be found. For example
   a. $f(x) = x^2 - 3x + 4$. In this case, the set R of all reals is taken as the domain of 'f'.
   b. $f(x) = \sqrt{16 - x^2}$ In order to have a real valued function, $16 - x^2 \geq 0$ i.e. $-4 \leq x \leq 4$ which the desired domain of 'f
   c. $f(x) = \dfrac{x+7}{x-7}$ then for $x = 4$, $x + 7$ is not defined. Hence in this case, the domain is the $x - 4$ set R of reals except 4 i.e. $R - \{4\}$

Example: If A is the area of the circle. Describe this function.

Solution: If we call this function f then r is the independent variable and A becomes the dependent variable and we have $A = f(r) = \pi r^2$ for $r \geq 0$. The domain must be stated with the constraint $r \geq 0$ as we can't have a circle with negative radius.

Example: A relation is given as:

$$\{ (1, 1), (2, \frac{1}{2}), (3, \frac{1}{3}), (4, \tfrac{1}{4}), (5, \frac{1}{5})\}$$

is it a function ? Justify. If it is a function state its domain and range. Exhibit the relation by a formula.

Solution: Let,

$$A = \{ 1, 2, 3, 4, 5\} \text{ and } B = \{1, \frac{1}{2}, \frac{1}{3}, \frac{1}{4}, \frac{1}{5}\}$$

Every element in *A* occurs once and only once in the ordered pairs and is related to one and only one element of the set *B*, exhibiting a relation.

Hence it is a function from the set *A* to the set *B*. Let us denote it by f. Then domain of *A* = The set A and the range of *f* = The set *B*. Also, the second member in each ordered pair is the reciprocal of the first member.

Thus if we denote any element of the set *A* by *x* and that of *B* by *y* then The formula is $f(x) = 1/x$ i.e. $y = 1/x$ $x \in A$.

Example: A function *f* is given as $f(x) = x^2 - 6x + 5$. State its domain.

Find $f(-2), f(2), f(1), f(a+h), \dfrac{f(a+h)-f(a)}{h}, h \neq 0$

Solution: The domain of '*f*' is the set R of all reals

$$f(x) = x^2 - 6x + 5 \Rightarrow f(-2) = (-2)^2 - 6(-2) + 5 = 21$$

$$f(2) = (2)^2 - 6(2) + 5 = -3, f(1)$$

$$= (1)^2 - 6(1) + 5 = 0$$

$$f(a+h) = (a+h)^2 - 6(a+h) + 5$$

$$= a^2 + 2ah + h^2 - 6a - 6h + 5$$

and,

$$= \frac{f(a+h)-f(a)}{h}$$

$$= \frac{(a^2+2ah+h^2-6a-6h+5)-(a^2-6h+5)}{h}$$

$$= \frac{h(2a-6+h)}{h}$$

Example: Find the range of each of the following functions:

1. $f(x) = 3x + 4$ for $-5 \leq x \leq 8$
2. $g(x) = 2x^2 + 3$ for $-4 \leq x \leq 3$
3. $\phi(x) = x^2 - 6x + 7$ for all $x \in R$

Solution:

1. The domain of ' f ' = $\{ x \mid x \in R, -5 \leq x \leq 8\}$

Now $-5 \leq x \leq 8 \backslash 3(-5) \leq 3(x) \leq 3(8)$

$\therefore -15 \leq 3x \leq 24 \therefore -15+4 \leq 3x+4 \leq 24+4$

$\therefore -11 \leq f(x) \leq 28$

The range of 'f' = $\{f(x) \in R, -11 \leq f(x) \leq 28\}$

2. $-4 \leq x \leq 3 \Rightarrow 0 \leq x^2 \leq 16$

$\therefore 2(0) \leq 2(x) \leq 2(16) \therefore 0+3 \leq 2x^2+3 \leq 30+3$

$\therefore 3 \leq 9(x) \leq 35$

This is the range of 'g'

3. $h(x)= x^2 - 6x + 7 = x^2 - 6x + 9 - 9 + 7 = (x-3)^2 - 2$

Since $x \in R$, $(x-3) \in R \Rightarrow (x-3)2 \geq 0 \Rightarrow (x-5)$ $2-2 \geq 0-2$

$\therefore h(x) \geq -2$. This is the range of 'h'.

Example: Find the largest possible set of real numbers which can be the domain of

$$f(x) = \sqrt{1-1/x}$$

Solution: Since,

$$\sqrt{1-1/x}$$

is defined only when,

$$\sqrt{1-1/x} \geq 0$$

i.e. $f(x)$ is defined for,

$$1-1/x \geq 0$$

i.e. $$f(x) = \frac{x-1}{x} \geq 0$$

Setting numerator and denominator separately equal to zero and solving, we get $x = 0$ and $x = 1$ (cut points)

Interval	Test value	Value of f (x)	Sign of f (x)
$(-\infty, 0)$	-1	$+2$	+
$(0, 1)$	$\frac{1}{2}$	$-\frac{1}{2}\cdot\frac{1}{2} = -1$	–
$(1, \infty)$	2	$\frac{1}{2}$	+

Graph of expression of $f(x) > 0$

Graph of expression of $f(x) \geq 0$

$\therefore$ The domain is the set $= \{x \mid x < 0 \text{ or } x \geq 1\} = (-\infty, 0) \cup [1, \infty)$

Example: If $f(x) = \dfrac{x+5}{3x-4}$ and $t = \dfrac{5+4x}{3x-1}$

then show that $f(t) = x$

Solution:

$$f(x) = \frac{x+5}{3x-4} \backslash f(t) = \frac{t+5}{3t-4} \text{ putting } t = \frac{5+4x}{3x-1}$$

$$\therefore f(x) = \frac{\frac{5+4x}{3x-1}+5}{3\left(\frac{5+4x}{3x-1}\right)}$$

$$= \frac{\frac{5+4x+15x-5}{3x-1}}{\frac{5+12x-12x+4}{3x-1}}$$

$$\therefore f(t) = \frac{19x}{19} = x \backslash f(t) = x$$

Example: If $f(x) = x^2 - 2x + 3$, $x \in R$. Find the values of x for which $f(x) = f(3x - 1)$

Solution:

$$f(x) = x^2 - 2x + 3 \Rightarrow f(3x-1)$$

$$= (3x-1)^2 - 2(3x-1) + 3$$

$$= 9x^2 - 6x + 1 - 6x + 2 + 3$$
$$= 9x^2 - 12x + 6$$
$$\therefore f(x) = f(3x - 1) \text{ gives us}$$
$$x^2 - 2x + 3 = 9x^2 - 12x + 6$$
$$8x^2 - 10x + 3 = 0$$
$$(4x - 3)(2x - 1) = 0$$
$$4x - 3 = 0 \text{ or } 2x - 1 = 0$$
$$\therefore x = 3/4 \text{ or } x = 1/2$$

$\therefore$ The required values of x are 3/4 and 1/2

CONCEPT OF LIMIT

Convergence

If an infinite sequence of numbers $a_1, a_2, a_3, \ldots, a_n$, is given and if there is a number l such that every interval, however small, marked off about the point l, contains all the points a_n, except for at most a finite number, we say that the number l is the limit of the sequence $a_1, a_2, \ldots$ or that the sequence $a_1, a_2, \ldots$ converges to l; in symbols, $\lim\limits_{x \to \infty} a_n = l$. Here we expressly remark that this includes the trivial case in which all the numbers a_n are equal to one another, and hence also coincide with the limit.

Instead of the above, we may use the following equivalent statement: If any positive number ε be assigned – however small – one can find a whole number N=N(ε) such that from N onward (i.e., for $n >$ N(ε)), it is always true that $|a_n - l| < \varepsilon$.. Of course, it is as a rule true that the bound $N(\varepsilon)$ will have to be chosen larger and larger with smaller and smaller values of ε; in other words, $N(\varepsilon)$ will increase beyond all bounds as ε tends to 0.

It is important to remember that every convergent sequence is bounded, that is, there corresponds to every sequence $a_1, a_2, a_3, \ldots$, for which a limit l exists, a positive number M, independent of n, such that for all the terms a_n of the sequence the inequality $|a_n| < M$ is valid. This theorem readily follows from our definition. We choose ε equal to 1; then there is an index N such that $|a_n - l| < 1$ for $n > N$. Let A be the largest among the numbers

$$|a_1 - l|, |a_2 - l|, \ldots, |a_n - l|$$

We can then put $M = [l] + A + 1$. Since, by the definition of A, the inequality $|a_n - l| < A + 1$ certainly holds for n= 1, 2, …, N, while for $n > N$

$$|a_1 - l| < 1 \leqq A + 1$$

A sequence which does not converge is said to be divergent. If, as n increases, the numbers a_n increase beyond all bounds, we say that the sequence diverges to +} and, as we have already done occasionally, we write $\lim_{n\to\infty} a_n = \infty$.

Similarly, we write $\lim_{n\to\infty} a_n = -\infty$, if, as n increases, the numbers $-a_n$ increase beyond all bounds in the positive direction. But divergence may manifest itself in other ways, as, for example, in the case of the sequence $a_1 = -1, a_2 = +1, \; a_3 = -1, a_4 = +1, \ldots$, the terms of which oscillate between two different values.

As regards convergence, the behaviour of a sequence is unaltered, if we omit a. finite number of the terms a_n. In what follows, we shall frequently use this, speaking of the convergence or divergence of series in which the term a_n is undefined for a finite number of values of n. In all the examples given above, the limit of the sequence considered is a known number.

If the concept of limit were to yield nothing more than the recognition that certain known numbers can be approximated as closely as we like by certain sequences of other known numbers, we should have gained very little. The fruitfulness of the concept of limit in analysis rests essentially on the fact that limits of sequences of known numbers provide a means of dealing with other numbers, which are not directly known or expressible.

The whole of higher analysis consists of a succession of examples of this fact. The representation of the irrational numbers as limits of rational numbers may be regarded as a first example. In this section, we shall become acquainted with further examples. However, before we take up this subject, we shall make a few general, preliminary remarks.

Intrinsic Convergence

How can we tell that a given sequence of numbers $a_1, a_2, a_3, \ldots, a_n$, converges to a limit, even when we do not know beforehand what that limit is? This important question is answered by Cauchy's convergence test. It is sometimes referred to as the general principle of convergence.

We say that a sequence of numbers $a_1, a_2, a_3, \ldots, a_n$, is convergent if there corresponds to every arbitrarily selected small positive number ε a number $N = N(\varepsilon)$, which usually depends on ε, such that $|a_n - a_m| < \varepsilon$, provided that n and m are both at least equal to $N(\varepsilon)$. Cauchy's convergence test can then be expressed as follows:

Every intrinsically convergent sequence of numbers possesses a limit. The importance of Cauchy's test lies in the fact that it allows to speak of the limit of a sequence after considering the sequence itself without any further information about the limit itself. The converse of Cauchy's test is very easy to prove.

For, if the sequence $a_1, a_2, \ldots$ tends to the limit l, then, by the definition of convergence, we have $|l - a_n| < \frac{\in}{2}$ and $|l - a_m| < \frac{\in}{2}$, where ε is a positive quantity as small as we please, provided only that both m and n are large enough, whence

$$|a_n - a_m| = |(l - a_m) - (l - a_n)| \leq |l - a_m| + |l - a_n| < \in$$

Since ε can be chosen as small as we please, this inequality expresses our statement.

Cauchy's test itself becomes intuitively obvious, if we think of the numbers as they are represented on the number axis. It then states that a sequence certainly has a limit, if after a certain point N all the terms of the sequence are restricted to an interval which can be made arbitrarily small by choosing N large enough.

Monotonic Sequence

The question whether a given sequence converges to a limit is particularly easy to answer when the sequence is a so-called monotonic sequence, that is, if either every number of the sequence is larger than the preceding number (monotonic increasing) or else every number is smaller than the preceding number (monotonic decreasing). We have the theorem:

Every monotonic increasing sequence the terms of which are bounded above (that is, lie below a fixed number) possesses a limit; similarly, every monotonic decreasing sequence the terms of which never drop below a certain fixed bound possesses a limit. Naturally, a convergent, monotonic, increasing sequence must tend to a limit which is greater than any term of the sequence, while for a convergent monotonic decreasing sequence the numbers tend to a limit which is smaller than any number of the sequence.

Thus, for example, the numbers $1/n$ form a monotonic decreasing sequence with the limit 0, while the numbers $1 - 1/n$ form a monotonic increasing sequence with the limit 1. In many cases, it is convenient to replace the condition that a sequence shall increase monotonically by the weaker condition that its terms shall never decrease, in other words, to allow its successive terms to be equal to one another. We then speak of a monotonic non-decreasing sequence, or of a monotonic increasing sequence in the wider sense. Our theorem on limits remains true for such sequences as well as for sequences which are monotonic non-increasing or monotonic decreasing in the wider sense.

LIMIT WITH VARIABLE CONTINUOUS

Hitherto, we have considered limits of sequences, that is, of functions of an integral variable n. However, the notion of limit frequently occurs in connection with the concepts of a continuous variable x and a function.

We say that the value of the function $f(x)$ tends to a limit l as x tends to ξ, or in symbols

$$\lim_{x \to \xi} f(x) = l,$$

if all the values of the function $f(x)$, for which x lies near enough to ξ, differ arbitrarily little from l. Expressed more precisely, the condition is:

If an arbitrarily small positive quantity ε is assigned, we can mark off about ξ an interval $|x - \xi| < \delta$ so small that there applies for every point x in this interval different from ξ itself the inequality $|f(x) - l| < \varepsilon$.

We expressly exclude here the equality of x and ξ. This is done purely for reasons of expediency, so as to have the definition in a form more convenient for applications, e.g., in the case where the function $f(x)$ is not defined at the point ξ, although it is defined for all other points in a neighbourhood of ξ.

If our function is defined or considered in a given interval only, for example, $\sqrt{1-x^2}$ in $-1 \leqq x \leqq 1$, we shall restrict the values of x to this interval. Thus, if ξ denotes an end-point of an interval, x is made to approach ξ by values on one side of ξ only (limit from the interior of the interval or one-sided limit).

As an immediate consequence of this definition, we have the fact:

If $\lim_{x \to \xi} f(x) = l$ and $x_1, x_2, \ldots, x_n$, is a sequence of numbers all different from ξ, but approaching ξ as a limit, then $\lim_{x \to \infty} f(x_n) = l$.

In fact, let ε be any positive number; we wish to show that for all values of n greater than a certain n_0 there applies the inequality.

By definition, there exists a $\delta > 0$ such that, whenever $|x - \xi| < \delta$, one has the inequality $|f(x_n) - l| < \in$.

Since $x_n < \xi$, the relation $|x - \xi| < \delta$ is satisfied for all sufficiently large values of n, and it follows for such values that $|f(x_n) - l| < \varepsilon$, as was to be proved. We shall now attempt to clarify this abstract definition by means of simple examples. Consider first the function

$$f(x) = \frac{\sin x}{x}$$

defined for x 0. We state that

$$\lim_{x \to 0} \frac{\sin x}{x} = 1$$

We cannot prove this statement simply by carrying out the

passage to the limit in the numerator and denominator separately, because the numerator and denominator vanish when $x = 0$, and the symbol 0/0 has no meaning. We arrive at the proof as follows:

From Fig, we find by comparing the areas of the triangles OAB and OAC and the sector OAB that, if $0 < x < \pi/2$,

$$\sin x < x < \tan x \cdot$$

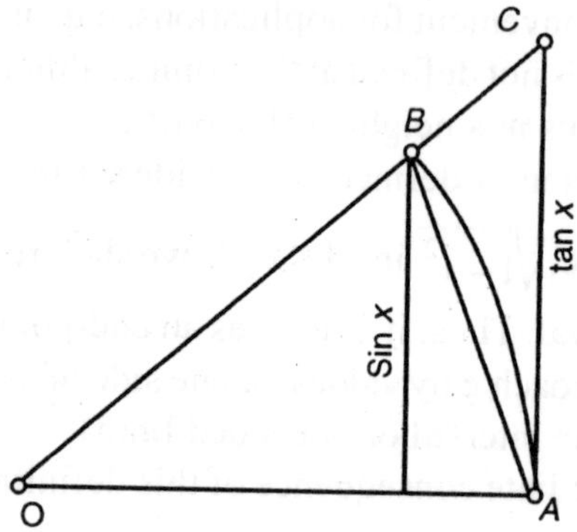

From this follows that, if $0 < |x| < \pi/2$,

$$1 < \frac{x}{\sin x} < \frac{1}{\cos x}$$

Hence the quotient $\sin x/x$ lies between the numbers 1 and $\cos x$. We know that $\cos x < 1$ as $x < 0$, whence the quotient $\sin x/x$ can only differ arbitrarily little from 1, provided that x is near enough to 0. This is exactly what is meant by the equation which was to be proved. By the result just proved

$$\lim_{x\to 0}\frac{\tan x}{x} = \lim_{x\to 0}\frac{\sin x}{x}\lim_{x\to 0}\frac{1}{\cos x} = 1,$$

and also $\quad \lim_{x\to 0}\dfrac{1-\cos x}{x} = 0$

This last result follows from the formula, valid for $0 < |x| < \pi/2$,

$$\frac{1-\cos x}{x} = \frac{(1-\cos x)(1+\cos x)}{x(1+\cos x)} = \frac{1-\cos^2 x}{x(1+\cos x)}$$

$$= \frac{\sin x}{x}.\frac{1}{1+\cos x}.\sin x.$$

As $x < 0$, the first factor on the right hand side tends to 1, the second one to 1/2 and the third one to 0, whence the product tends to 0, as has been stated.

The same formula, on division by x, yields whence

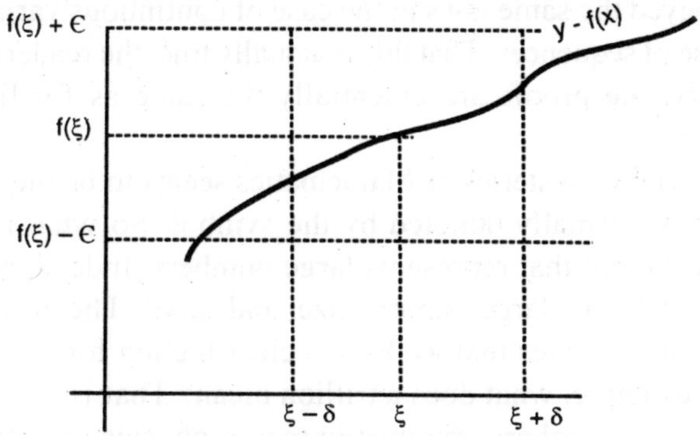

$$\frac{1-\cos x}{x^2} = \left(\frac{\sin x}{x}\right)^2 \frac{1}{1+\cos x}$$

$$\lim_{x\to 0} \frac{1-\cos x}{x^2} = \frac{1}{2}$$

Finally, consider the function $\sqrt{x^2}$, defined for all values of x. This function is never negative; it is equal to x for $x > 0$ and to $-x$ for $x < 0$. In other words $\sqrt{x^2} = |x|$.

Hence the function $\sqrt{x^2}/x$, defined for all non-zero values of x, has the value $+1$ when $x > 0$ and -1 when $x < 0$, whence the limit cannot exist, since arbitrarily near to 0 we can find values of x for which the quotient is $+1$ and other values for which it is -1.

In concluding this discussion on limits in connection with a continuous variable, we remark that it is, of course, possible to consider limiting processes in which the continuous variable x increases beyond all bounds. We state the example

$$\lim_{x\to\infty} \frac{x^2+1}{x^2-1} = \lim_{x\to\infty} \frac{1+\frac{1}{x^2}}{1-\frac{1}{x^2}} = 1$$

without further discussion. It signifies that the function on the left hand side differs arbitrarily little from 1, provided only that x is sufficiently large.

In these examples, we have proceeded as if operations with limits obeyed the same laws in the case of continuous variables as in the case of sequences. That this is actually true, the reader himself can verify; the proofs are essentially the same as for limits of sequences.

One of the mysteries of Mathematics seems to be the concept of "infinity", usually denoted by the symbol. So what is ? It is simply a symbol that represents large numbers. Indeed, numbers are of three kinds: large, normal size, and small. The normal size numbers are the ones that we have a clear feeling for.

For example, what does a trillion mean? That is a very large number. Also numbers involved in macro-physics are very large numbers. Small numbers are usually used in micro-physics. Numbers like 10^{-75} are very small. Being positive or negative has special meaning depending on the problem at hand. The common mistake is to say that $-\infty$ is smaller than 0. While this may be true according to the natural order on the real line in term of sizes, $-\infty$ is big, very big!

So when do we have to deal with and $-\infty$? Easy: whenever you take the inverse of small numbers, you generate large numbers and vice-versa. Mathematically we can write this as:

$$\frac{1}{0} = \pm\infty \text{ and } \frac{1}{\infty} = 0$$

Note that the inverse of a small number is a large number. So size-wise there is no problem. But we have to be careful about the positive or negative sign. We have to make sure we know whether a small number is positive or negative. 0+ represents small positive numbers while 0- represents small negative numbers. (Similarly, we will use e.g. 3+ to denote numbers slightly bigger than 3, and 3- to denote numbers slightly smaller than 3.) In other words, being more precise we have

$$\frac{1}{0+} = \pm\infty \text{ and } \frac{1}{0-} = -\infty$$

Remark: Do not $\pm\infty$ treat as ordinary numbers. These symbols do not obey the usual rules of arithmetic, for instance $\infty + 1 = \infty, \infty - 1 = \infty, 2, \infty = \infty$, etc.

Example: Consider the function

$$f(x) = \frac{2x+1}{x-1}$$

We have

$$\frac{2x+1}{x-1} = \frac{2+\frac{1}{x}}{1-\frac{1}{x}}$$

which implies

$$\lim_{x\to\pm\infty} f(x) = \frac{2+0}{1-0} = 2.$$

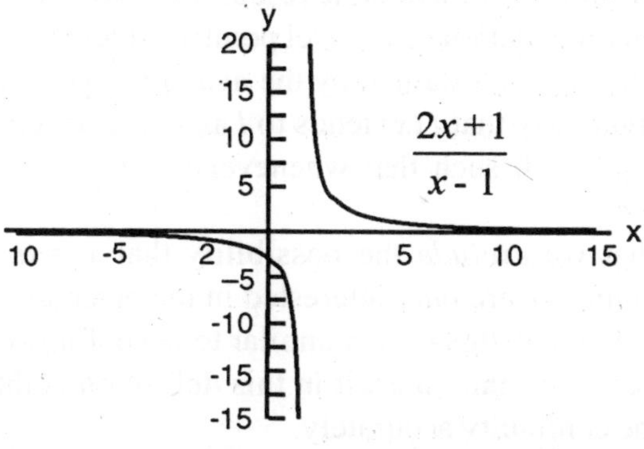

Note that when x gets closer to $\pm\infty$ (x gets large), then the points on the graph get closer to the horizontal line $y = 2$. Such a line is called a horizontal asymptote. In particular, we have

$$\lim_{x\to-\infty} \frac{a}{x^r} = 0 \; and \; \lim_{x\to+\infty} \frac{a}{x^r} = 0$$

for any number a, and any positive number r, provided x^r is defined. We also have

$$\lim_{x\to\infty} x^r = \infty$$

For $-\infty$, we have to be careful about the definition of the power of negative numbers. In particular, we have

$$\lim_{x \to -\infty} x^n = (-1)^n \infty$$

for any natural number *n*.

Many things grow at different rates which can be described by functions. Yet each of these functions of growth has some factor that allows it to grow only so much. An often used example is population growth. Populations (of people, animals, insects, bacteria, etc.) can continue to grow only as long as they have enough room and food. If the amount of food and space to live are constant, then as the population increases, there will be less for each individual to consume and less room to live and grow.

When the population exceeds a size that can survive with the given food and space it grows too big it will begin to decrease due to death. The population will level out around a number that can be sustained with the available resources, and stay around that number. Every population has a level point at which it can survive; a number that can be sustained by the available space, a limit.

Definition: Say that $f(x)$ tends to l as $x \to a$ iff given $\in > 0$, there is some $\delta > 0$ such that whenever $0 < |x - a| < \delta$, then $|f(x) - l| < \in$.

Note that we *exclude* the possibility that $x = a$ when we consider a limit; we are *only* interested in the behaviour of f near a, but not at a. In fact this is very similar to the definition we used for sequences. Our main interest in this definition is that we can now describe continuity accurately.

***Definition*:** Say that f is continuous at a if $\lim_{x \to \alpha} f(x) = f(a)$. Equivalently, f is continuous at a iff given $\in > 0$, there is some $\delta > 0$ such that whenever $|x - a| < \delta$, then $|f(x) - f(a)| < \in$.

Note that in the "epsilon – delta" definition, we no longer need exclude the case when $x = a$. Note also there is a good geometrical meaning to the definition. Given an error, there is *some* neighbourhood of a such that if we stay in that neighbourhood, then f is trapped within of its value $f(a)$. We shall not insist on this definition in the same way that the definition of the convergence of a sequence was emphasised. However, all our work on limts and continuity of functions can be traced back to this definition, just as in our work on sequences, everything could be traced back to the definition of a convergent sequence.

Rather than do this, we shall state without proof a number of results which enable continuous functions both to be recognised and manipulated.

So you are expected to know the definition, and a few simply $\in -\delta$ proofs, but you can apply (correctly – and always after checking that any needed conditions are satisfied) the standard results we are about to quote in order to do required manipulations.

Definition: Say that f: $U(open) \to \mathbb{R}$ is continuous if it is continuous at each point $a \in U$.

Note:This is important. The function $f(x) = 1/x$ is defined on $\{x{:}x \neq 0\}$, and is a continuous function. We cannot usefully define it on a larger domain, and so, by the definition, it is continuous. This is an example where the naive "can draw it without taking the pencil from the paper" definition of continuity is not helpful.

Example: Let $f(x)\dfrac{x^3-8}{x-2}=$ for $x \neq 2$. Show how to define f (2) in order to make f a continuous function at 2.

Solution: We have

$$\frac{x^3-8}{x-2} = \frac{(x-2)\left(x^2+2x+4\right)}{(x-2)} = (x^2 + 2x + 4)$$

Thus $f(x) \to (2^2 + 2.2 + 4) = 12$ as $x \to 2$. So defining $f(2) =$ 12 makes f continuous at 2, (and hence for all values of x).

Operations with Limits

We conclude with a remark concerning calculations with limits. It follows almost at once from the definition of limit that we can perform the elementary operations of addition, multiplication, subtraction, and division according to the rules:

If $a_1, a_2,...$ is a sequence with the limit a and $b_1, b_2,...$ a sequence with the limit b, then the sequence of numbers $c_n = a_n + b_n$ also has a limit and

$$\lim_{n\to\infty} c_n = a + b$$

The sequence of numbers $c_n = a_n b_n$ likewise converges, and

$$\lim_{n\to\infty} c_n = ab$$

Similarly, the sequence $c_n = a_n - b_n$ converges, and

$$\lim_{n\to\infty} c_n = a - b$$

Provided the limit b differs from 0, the numbers $c_n = a_n/b_n$ also converge, and have the limit

$$\lim_{n\to\infty} c_n = \frac{a}{b}$$

In words: we can interchange the rational operations of calculation with the process of forming a limit, that is, we obtain the same result whether we first perform a passage to the limit and then a rational operation or *vice versa*.

For the proof of these simple rules, it is sufficient to give one example; using this as a model, the reader can establish the other statements by himself.

As this example, we consider multiplication of limits. The relations $a_n \to a$ and $b_n \to b$ amount to the following: If we choose any positive number ε, we need only take n greater than N, where $N = N(\varepsilon)$ is a sufficiently large number depending on ε, in order to have both

$$|1 - a_n| < \in \text{ and } |b - b_n| < \in.$$

If we write $ab - a_n b_n = b(a - a_n) + a_n(b - b_n)$ and recall that there is a positive bound M, independent of n, such that $|a_n| < M$, we obtain

$$|ab - a_n b_n| \leqq |b||a - a_n| + |a_n||b - b_n| < (|b| + M)\in$$

Since the quantity $(|b| + M)\varepsilon$ can be made arbitrarily small by choosing ε small enough, we see that the difference between ab and $a_n b_n$ actually becomes as small as we please for all sufficiently large values of n, which is precisely the statement made in the equation

$$ab = \lim_{n\to\infty} a_n b_n$$

By means of these rules, many limits can be evaluated very easily; for example, we have

$$\lim_{n\to\infty} \frac{n^2 - 1}{n^2 + n + 1} = \lim_{n\to\infty} \frac{1 - \frac{1}{n^2}}{1 + \frac{1}{n} + \frac{1}{n^2}} = 1$$

since in the second expression the passage to the limit in the numerator and denominator can be made directly. Another simple and obvious rule is worth stating. If $\lim a_n = a$ and $\lim b_n = b$, and if, in addition, $a_n > b_n$ for every n, then $a \geq b$. However, we are by no means entitled to expect that, in general, a will be greater than b, as is shown by the case of the sequences $a_n = 1/n$, $b_n = 1/2n$, for which $a = 0 = b$.

The Number '*e*'

As a first example of the generation of a number, which cannot be stated in advance as the limit of a sequence of known numbers, we consider the sums

$$S_n = 1 + \frac{1}{1!} + \frac{1}{2!} + \ldots + \frac{1}{n!}.$$

We assert that, as n increases, S_n tends to a definite limit.

In order to prove the existence of the limit, we observe that, as n increases, the numbers S_n increase monotonically. For all values of n, we also have

$$S_n \leq 1 + 1 + \frac{1}{2} + \frac{1}{2^2} + \ldots + \frac{1}{2^{n-1}} = 1 + \frac{1 - \frac{1}{2^n}}{1 - \frac{1}{2}} < 3.$$

Hence the numbers S_n have the upper bound 3 and, being a

monotonic increasing sequence, they possess a limit, which we denote by e:

$$e = \lim_{n\to\infty} S_n.$$

Moreover, we assert that the number e, defined as the above limit, is also the limit of the sequence

$$T_n = \left(1+\frac{1}{n}\right)^n$$

The proof is simple and at the same time an instructive example of operations with limits. According to the binomial theorem, which we shall here assume,

$$\begin{aligned}
T_n &= \left(1+\frac{1}{n}\right)^n \\
&= 1+n\frac{1}{n}+\frac{n(n-1)}{2!}\frac{1}{n^2}+\ldots+\frac{n(n-1)(n-2)\ldots 1}{n!}\frac{1}{n^n} \\
&= 1+1+\frac{1}{2!}\left(1-\frac{1}{n}\right)+\ldots \\
&= +\frac{1}{n!}\left(1-\frac{1}{n}\right)\left(1-\frac{2}{n}\right)\ldots\left(1-\frac{n-1}{n}\right).
\end{aligned}$$

Hence we see at once: (1) that $T_n \geq S_n$, and the T_n also form a monotonic increasing sequence*, whence there follows the existence of the limit $\lim_{n\to\infty} T_n = T$. In order to prove that $T = e$, we observe that

$$T_m > 1+1+\frac{1}{2!}\left(1-\frac{1}{m}\right)+\ldots+\frac{1}{n!}\left(1-\frac{1}{m}\right)\ldots\left(1-\frac{n-1}{m}\right),$$

provided that $m > n$. If we now keep n fixed and let m increase beyond all bounds, we obtain on the left hand side the number T and on the right hand side the expression S_n, so that $T S_n$.

We have thus established the relationship $T \geq S_n \geq T_n$ for every value of n. We can now let n increase, so that T_n tends to T; it follows from the double inequality that $T = \lim_{n\to\infty} S_n = e$ as was to be proved.

The number π as a limit: A limiting process, which in essence goes back to classical antiquity (Archimedes), is that by which the number π is defined. Geometrically speaking, π means the area of the circle of radius 1.

We therefore accept the existence of this number as intuitive, regarding it as obvious that this area can be expressed by a (rational or irrational) number, which we then simply denote by π. However, this definition is not of much help to us, if we wish to calculate the number with any accuracy.

We have then no choice but to represent it by means of a limiting process, namely, as the limit of a sequence of known and easily calculated numbers.

Archimedes himself used this process in his method of exhaustions, where he steadily approximated the circle by means of regular polygons with an increasing number of sides fitting it more and more closely. If we denote by f the area of the regular polygon with m sides, inscribed in the circle, then the area of the inscribed $2m$ polygon is given by the formula.

$$\frac{m}{2}\sqrt{2-2\sqrt{1-\left(\frac{2f_m}{m}\right)^2}}$$

We now let m run, not through the sequence of all positive integers, but through the sequence of powers of 2, that is, $m = 2^n$; in other words, we form those regular polygons the vertices of which are obtained by repeated bisection of the circumference. The area of the circle is then given by the limit:

$$\pi = \lim_{n\to\infty} f_{2^n}$$

In fact, this representation of π as a limit serves as a base for numerical computations; starting with the value $f_4 = 2$, we can calculate in order the terms of our sequence tending to π. An estimate of the accuracy with which any term f_{2^n} represents π can be obtained by constructing the lines touching the circle and parallel to the sides of the inscribed 2^n polygon. These lines form a circumscribed polygon, similar to the inscribed 2^n polygon, and have greater dimensions in the ratio $1 : \cos\pi/2^{n-1}$, whence the area F_{2^n} of the circumscribed polygon is given by

$$\frac{f_{2^n}}{F_{2^n}} = \left(\cos\frac{\pi}{2^{n-1}}\right)^2.$$

Since the area of the circumscribed polygon is evidently larger than that of the circle, we have

$$f_{2^n} < \pi < F_{2^n} = \lim_{x\to\infty} f(x_n) = l$$

These are matters with which the reader will be more or less familiar. What we wish to point out here is that the calculation of areas by means of exhaustion by rectilinear figures, the areas of which are readily calculated, forms the basis for the concept of integral to be introduced.

6

Derivative

Calculus involves analyzing instantaneous changes with reference to the entire system. We saw how using a small interval of $\Delta x = 2$ caused great differences in the rate of change of the function over the same interval. The function did not change nearly as much from $x = 0$ to $x = 2$ as it did from $x = 4$ to $x = 6$.

The rate at which the function is changing must be dependent on the value of the function. In other words the rate of change must be defined at each interval and this value must be unique to that interval.

We can approximate the graph from $x = 6$ *to* $x = 8$ more closely by using a smaller interval of Δx=0.5. This will give an even more accurate description of the graph of the function $f(x) = x^2$ over that interval. Keep in mind that the rate of change of the function varies with x, and the only way to accurately analyze this change is by breaking the graph up into smaller and smaller sub-intervals and study what is happening over these infinitely small intervals:

First let us define a method for calculating the rate of change between any two points on the graph of *f*(*x*)*:* If the change in *f* of a function from a point *x* to another point ($x + \Delta x$) is given by:

$$\Delta f = f(x+\Delta x) - f(x)$$

The rate of change of *f(x)* over this interval is found by dividing by Δx.

$$\frac{\Delta f}{\Delta x} = \frac{f(x+\Delta x) - f(x)}{\Delta x}$$

Remember the Δx on the left side is the same as the Δx in the $f(x + \Delta x)$ expression.

From $x = 6$ *to* $x = 6.5$ the rate of change is:

$$\frac{\Delta f}{\Delta x} = \frac{f(6+0.5) - f(6)}{0.5} = 12.5$$

From $x = 6.5$ to $x = 7$ the rate of change is:

$$\frac{\Delta f}{\Delta x} = \frac{f(6.5+0.5) - f(6.5)}{0.5} = 13.5$$

Similarly the rate of change *from* $x = 7$ *to* $x = 7.5$ *and* $x = 7.5$ *to* $x = 8$ is 14.5 and 15.5 respectively. Notice how the rate of change of the function *slightly* increases as x increases. This implies that there must be a way to instantaneously define the rate of change of function through "*one point*" on the graph.

Before continuing, take a look at how the average rate of changes or slopes just calculated, approximate the graph from $x = 6$ *to* $x = 8$.

As you can see from the above graph, the smaller interval of $\Delta x = 0.5$ gives a more accurate approximation of the graph of $f(x)$. Our results are only averages because the rate of change of $f(x)$ is increasing with x and is not constant over an interval Δx. To define the instantaneous rate of change of we need to define Δx as an infinitely small distance separating two point x and $(x+ \Delta x)$ on the graph of $f(x)$.

The rate of change calculated through these two points will then give us the exact rate of change of $f(x)$ at that point, x. Since through any one point there exists an infinite many lines, we are in actuality giving the slope of the line through the two points x and $(x+\Delta x)$ where Δx goes to zero. The two points are assumed to be infinitely close to each other such that the rate of change $f(x)$ over that interval is constant. As we let these two points (x) and $(x+ \Delta x)$ come closer and closer together, the slope of the line through these two points corresponds to the actual graph itself. For example with Δx=2, the line through $x = 6$ an d $x = 8$ only roughly represented the actual graph.

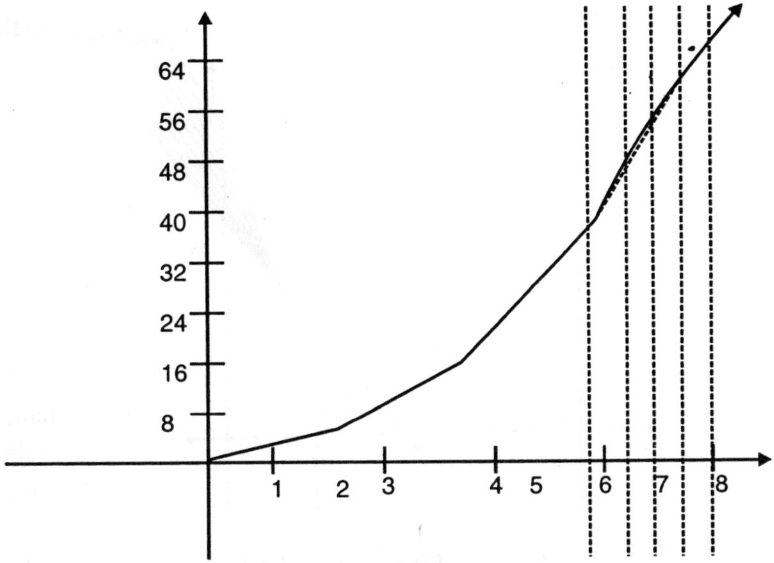

However by breaking up the interval from $x = 6$ to $x = 8$ into four sub-intervals of $\Delta x = 1/2$, we got a better approximation to the actual graph itself.

The instantaneous rate of change can now be defined as the rate of change of the line through a point x and $(x + \Delta x)$ where Δx is near to zero. This is called taking the limit as Δx goes to zero. In the graph this looks like.

By letting the difference or Δx between the two points go to zero we are able to accurately describe the behavior of the function at any *point* on the graph. With Δx large we can only approximate the functions behavior over the interval.

This infinitely small Δx allows us to define the rate of change of $f(x)$ through any one point on the graph, where point is by definition an infinitely small interval between x and $x + \Delta x$ over which the rate of change is assumed to be constant or does not change.

Thus the instantaneous rate of change can also be thought of as the tangent to the graph at any point, where point is actually two simultaneous points on the graph, separated by an infinitely small distance. For example the tangent to the graph at $x = 6$ is given by:

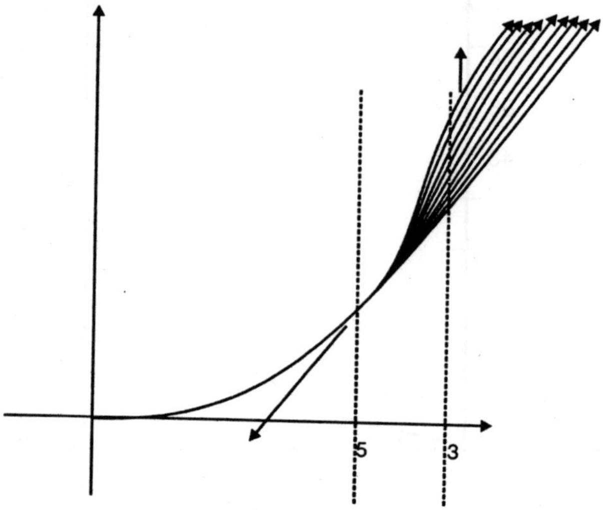

$$\frac{\Delta f}{\Delta x} = \lim_{\Delta x \to 0} \frac{(6+\Delta x) - f(x)}{\Delta x}$$

$$\frac{\Delta f}{\Delta x} = \lim_{\Delta x \to 0} \frac{(6+\Delta x)^2 - (6)^2}{\Delta x}$$

$$\frac{\Delta f}{\Delta x} = \lim_{\Delta x \to 0} \frac{36 + 12\Delta x + \Delta x^2 - 36}{\Delta x}$$

$$\frac{\Delta f}{\Delta x} = \lim_{\Delta x \to 0} 12 + \Delta x$$

$$\frac{\Delta f}{\Delta x} = \lim_{\Delta x \to 0} 12 + 0 = 12$$

We can now go on to define the rate of change of the at any point x of the function $f(x) = x^2$ as:

$$\frac{\Delta f}{\Delta x} = \lim_{\Delta x \to 0} \frac{f(x+\Delta x) - f(x)}{\Delta x}$$

$$\frac{\Delta f}{\Delta x} = \lim_{\Delta x \to 0} 2x + \Delta x$$

$$\frac{\Delta f}{\Delta x} = 2x$$

This function is called the derivative and gives us the rate of change or tangent to graph of $f(x) = x^n$ at any point x. It is extremely important to realize that the function's derivative is also a function of x. In other words its value changes with x to reflect a change in the rate of change of the original function. This may sound a bit confusing but I encourage you to spend a few minutes and think how it is so. Since the function's rate of change is changing then it can only be constant over an infinitely small interval or point. It is through this point that we define the derivative by taking the limit as Δx goes to zero.

In effect we define the graph of the function through a series of connected lines with changing slopes.

In the following paragraph we will mathematically prove that the derivatives value converges to the exact value as all errors go to zero as Δx goes to zero. Since the graph of the function is continuous over an infinitely small interval then so will its derivative be defined for that interval or point. Note that Δx and Δf no longer refer to discrete values. When we let Δx go to zero, its value becomes infinitely small such that:

$$\frac{\Delta f}{dx} = \lim_{\Delta x \to 0} \frac{\Delta f}{\Delta x}$$

Here *df* and *dx* represent an infinitely small difference in *x*, where:

$$dx \to 0$$

We'll now see how to find the derivative of any simple function, $f(x) = x^n$. Since the derivative is also a function , we shall call it $f'(x)$, "f prime of x" whe

$$f'(x) = \frac{df}{dx}$$

Our results can now be generalized for any function $f(x) = x^n$ where n is a positive integer.

$$f'(x) = \frac{df}{dx} = \lim_{\Delta x \to 0} \frac{f(x + \Delta x) - f(x)}{\Delta x}$$

$$f'(x) = \lim_{\Delta x \to 0} \frac{(x + \Delta x)^n - (x)^n}{\Delta x}$$

From the Binomial expansion theorem we get:

$$f'(x) = \lim_{\Delta x \to 0} \frac{x^n + nx^{n-1}\Delta x + \frac{(n)(n-1)}{2!}x^{n-2}\Delta x^2}{\Delta x}$$

$$\lim_{\Delta x \to 0} \frac{+\ldots\ldots \frac{(n)(n-1)(n-2)\ldots(n-(n-1))}{n!}x^0\Delta x^n - x^n}{\Delta x}$$

This simplifies to

$$f'(x) = \lim_{\Delta x \to 0} \frac{nx^{n-1}\Delta x + \frac{(n)(n-1)}{2}x^{n-2}\Delta x^2}{\Delta x}$$

$$\lim_{\Delta x \to 0} \frac{+\frac{(n)(n-1)(n-2)x^{n-3}\Delta x^{3n}}{3!}}{\Delta x}$$

We can replace all the

$$\frac{(n)(n-1)}{2!}\frac{(n)(n-1)(n-2)}{3!}$$

$c_1, c_2 \ldots . c_n c_1, c_2 \ldots . c_n$ are constants:

$$f'(x) = \lim_{\Delta x \to 0} \frac{nx^{n-1}\Delta x + c_1 x^{n-2}\Delta x^3 \ldots . + c_n x^0 \Delta^n}{\Delta x}$$

Take note of how x goes from x to n'th power to the 0 power while Δx goes from the 0 power to the n'th power in the last term. If we divide through by Δx we get:

$$f'(x) = \lim_{\Delta x \to 0} nx^{n-1} + c_1 x^{n-2}\Delta x + c_2 x^{n-3}\Delta x \ldots . . c_n x^0 \Delta x^n$$

As we take the limit as Δx goes to zero, every term after the first one goes to zero and hence cancels out; This leaves us with:

$$f'(x) = nx^{n-1} + 0 + +0 + 0\ldots$$

$$f'(x) = \frac{df}{dx} = n.x^{n-1}$$

The rate of change of a function $f(x) = x^n$ at any point x is given by the function's derivative $f'(x)$ evaluated at that point x.

RATE OF CHANGE

The derivative of a function at a chosen input value describes the best linear approximation of the function near that input value. For a real-valued function of a single real variable, the derivative at a point equals the slope of the tangent line to the graph of the function at that point. In higher dimensions, the derivative of a function at a point is a linear transformation called the linearization. The process of finding a derivative is called differentiation.

The fundamental theorem of calculus states that differentiation is the reverse process to integration. The mathematical relationship between an independent and dependent dimension. The relationship expressed in the form of a mathematical function defines a situation for a given set of conditions and properties. Since we allowed the independent dimension to change, it was no longer a fixed quantity but became a variable.

The graph of the function reflected a horizontal change of the independent variable with a vertical change in the dependent dimension. This graph allows us to visualize how a situation changes with respect to a change in the conditions. To find out how much the function, $f(x)$, changes between two points x_1 and x_2, we simply enter in the two values for the independent variable x and then calculate the difference between the dependent variable, *f*, for those given conditions.

Remember that a variable is nothing more than a dimension that is allowed to change or take on any value. Thus, from x_1 to x_2, the change in the independent variable , referred to as Δx is:

$$\Delta x = x_1 - x_2$$

The corresponding change in the dependent variable , referred to as Δf is:

$\Delta f = f(x_2) - f(x_1)\Delta x$ refers to an interval over which we are analyzing the change in the dependent dimension, *f*.. The second point x_2 can be written in terms of the first point x_1 plus the change in the variable, Δx

$$x_2 - x_1 = \Delta x$$
$$\therefore x_2 = x_1 + \Delta x$$

Therefore the change in the dependent variable over an interval $x_1 x_2$ from $\Delta f = f(x_2) - f(x_1)$ which can also be written as:

$$\Delta f = f(x_1 + \Delta x) - f(x_1)$$

where $\Delta x = x_2 - x_1$

For example the change in the function $f(x) = 3 \cdot x^3 x = 3$ to $x = 5$, where

Δx is $5 - 3 = 2$ is:

$$\Delta f = f(x_1 + \Delta x) - f(x_1)$$
$$\Delta f = f(3+2) - f(3)$$
$$\Delta f = 3.5^3 - 3.3^3 = 375 - 81$$
$$\Delta f = 294$$

AVERAGE RATE OF CHANGE

We have learned that a change in the independent variable is defined as $\Delta x = x_2 - x_1$, and the corresponding change in the dependent variable over this interval is $\Delta f = f(x_1 + \Delta x) - f(x_1)$ The question we now must ask ourselves is how can we measure the relative change of the dependent variable with respect to the independent variable? In other words how can we calculate how much more or less Δf changed compared to Δx.

To calculate how much more $f(x)$ changed over an interval from x_1, x_2, we simply divide the change in f over the change in *x* for the interval. Thus we divide, Δf by the interval over we which are evaluating it, Δx which is equal to $x_2 - x_1$. Thus the relative change of *f* with respect to *x* over an interval Δx is defined as:

$$\frac{\Delta f}{\Delta x} = \frac{f(x_2) - f(x_1)}{x_2 - x_1}$$

While this expression may seem rather simple, it does require some explanation. By dividing the change in *f* by the change in *x* what we are doing is calculating how much more f changed for a given change in x. For example in the function, $f(x)=3.x^3$, when x changed from 3 to 5, f changed from 81 to 375. Over this interval of from $x = 3$ to $x = 5$, the Δf was 294.

Thus the relative change in f with respect to a change in the independent variable x is:

$$\frac{\Delta f}{\Delta x}=\frac{f(x_2)-f(x_1)}{x_2-x_1}$$

The value of 147 tells us that f changes 147 times faster than x over that interval of Δx from $x = 3$ to $x = 5$ only. Thus for each unit change in x, $\Delta x = 1$, the corresponding change in f is 147. We can therefore define the rate of change of a function with respect to its independent variable to be:

$$\frac{\Delta f}{\Delta x}=\frac{f(x_1+\Delta x)-f(x_1)}{\Delta x}$$

The value, called the rate of change of the function, refers to how much more or less $f(x)$ changes for a unit change in x. Itis only valid is for the interval under consideration, $\Delta x = x_2 - x_1$.

Another way of understanding what rate of change of a function means is to look at the steepness of the line connecting the two endpoints of the interval under consideration.

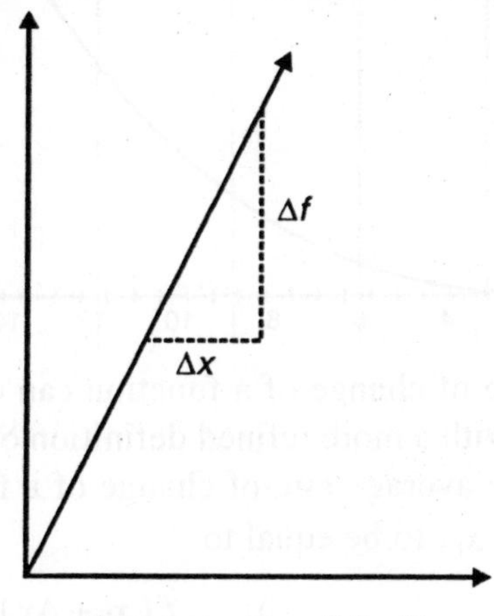

As Δf increases, the steepness of the line connecting the two endpoints will increase. Thus, the greater the rate of change of the function, the greater its slope or steepness over the interval under consideration.

Since slope and rate of change are synonymous, then how is rate of change defined for functions whose graphs do not have constant slopes? For example, from $x = 9$ to $x = 12$ of $f(x) = 3.x^3$, the change in f is 2997. Thus the rate of change of the function over the interval is:

$$\frac{\Delta f}{\Delta x} = \frac{2997}{3} = 999$$

This value is significantly higher than the rate of change calculated for the previous interval from $x = 3$ to $x = 5$. We can only conclude that the rate of change or slope of the graph must be increasing and is not constant over an interval $\Delta x = x_2 - x_1$. Look at the graph of the function $f(x) = 3.x^3$ to understand how this might be so:

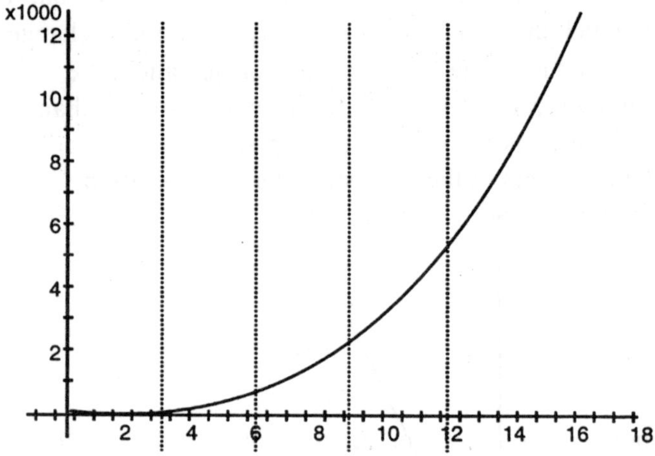

Since the rate of change of a function can change, then we have to come up with a more refined definition of rate of change. We can define the average rate of change of a function over an interval $\Delta x - x_2 - x_1$, to be equal to

$$\text{Average rate of change} = \frac{\Delta f}{\Delta x} = \frac{f(x_1 + \Delta x) - f(x_1)}{\Delta x}$$

In the next section we will take a closer look at how we can define the exact or instantaneous rate of change of a function.

INSTANTANEOUS RATE OF CHANGE

In this section we will take a much closer look at rate of change and see how we can define the instantaneous rate of change of a function at any point or value of the independent variable.

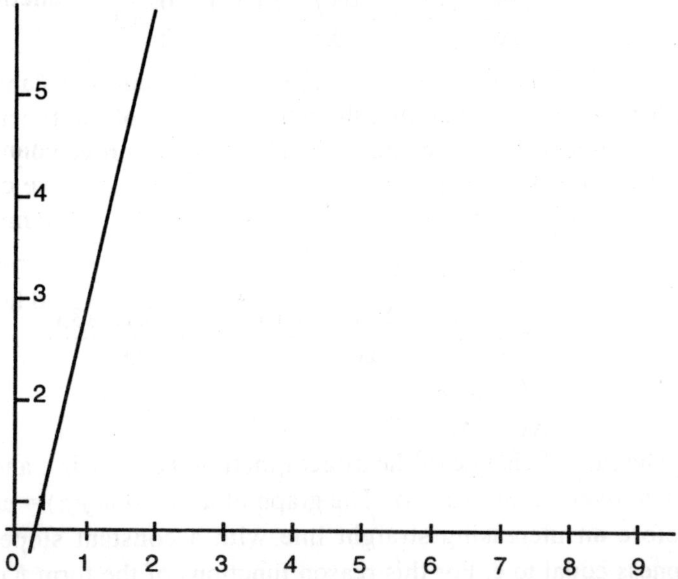

Let us begin with a study of the simple $f(x) = 3x$. Functions of the form $f(x) = cx$, where c is a constant, express direct relationships. This is because the value of the function, f, is a constant multiple or fraction of the independent variable. Thus f is said to be directly proportional to x. The graph of the function $f(x) = 3x$ looks like. From $x = 1$ to $x = 3$, $\Delta x = 2$, and $\Delta f = 9 - 3 = 6$. Thus, over this interval of $\Delta x = 2$, Δf equals 6. The average rate at which f changed with respect to x is by definition, $\frac{\Delta f}{\Delta x} = \frac{6}{2} = 3$. For each unit change in x, the change in f is 3. This tells us that f is changing three times faster that x is changing over the interval from $x = 1$ to $x = 3$. We can now look at the interval from $x = 2$ to $x = 4$ where Δx equals 2.

$$\Delta f = f(x_1 + \Delta x) - f(x_1)$$
$$\Delta f = f(2+2) - f(2)$$
$$\Delta f = 3.(4) - 3.(2)$$
$$\Delta f = 6$$

Thus

$$\frac{\Delta f}{\Delta x} = \frac{f(x_1 + \Delta x) - f(x_1)}{\Delta x} = \frac{6}{2} = 3$$

Over this interval, the rate of change is the same constant, 3. This leads us to conclude that the rate of change of the function over any interval is a constant, 3. This can be proven by the definition of rate of change:

$$f(x) = 3x$$
$$f(x + \Delta x) = 3(x + \Delta x)$$
$$\frac{\Delta f}{\Delta x} = \frac{f(x+\Delta x) - f(x)}{\Delta x} = \frac{3x + 3\Delta x - 3x}{\Delta x}$$
$$\frac{\Delta f}{\Delta x} = \frac{3\Delta x}{\Delta x} = 3$$

The rate of change of the direct function *f*(*x*) = *cx* is c and is constant over an interval, Δx. The graph of a function *f*(*x*) = *cx* is therefore an increasing straight line with a constant slope or steepness equal to c. For this reason functions of the form *f*(*x*) = *cx* are also called linear functions since their graphs are straight lines with a constant rate of change.

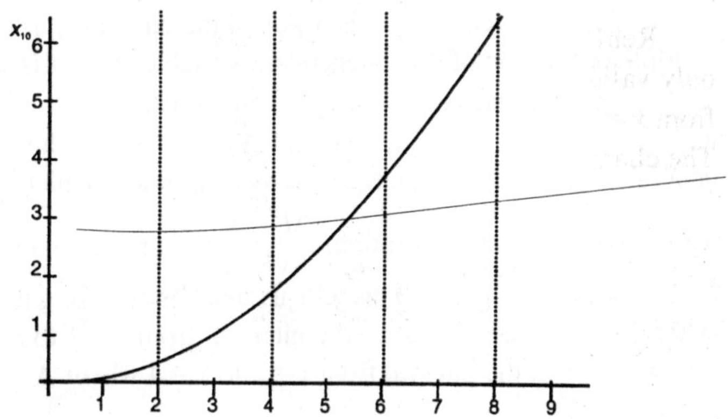

On the other hand, the graph of the function $f(x) = x^2$ is not a straight line. Unlike a line, the rate of change of $f(x)$ is not the same constant over any interval Δx. To define the rate of change for the function we will have to derive a more precise way of defining rate of change of function. To do this we will analyze the function over small intervals of Δx

The first point to notice is that the rate of change of the function varies with x. When x is small, f does not change that much as compared to how much it changes when x is large. We can conclude that the rate of change of f with respect to x, is not constant over any interval, Δx, but varies with x. To begin our analysis let us divide up the graph into intervals of $\Delta x = 2$ and study what is happening in each of these intervals separately. From $x = 0$ to $x = 2$, $\Delta x = 2$, the change in f, Δf, *is* $f(2) - f(0) = 4$. From $x = 6$ to $x = 8$, $\Delta x = 2$, the change in f, Δf, *is* $f(8) - f(6) = 28$.

Clearly, as x increases the rate of change of the function is increasing and is not a constant as in our study of the line where the function changed at the same rate as x increased.

Returning back to the graph of $f(x) = x^2$; we calculated that from $x = 0$ *to* $x = 2$ the change in f was 4. We can then conclude that from $x = 0$ to $x = 2$ the average rate of change of the function over that particular interval is 2 or:

$$\frac{\Delta f}{\Delta x} = \frac{4}{2} = 2$$

Remember it is called average because this rate of change is only valid from $x = 0$ to $x = 2$. Now let us consider the interval from $x = 2$ to $x = 4$ where once again the change in x, Δx, is 2. The change in f of the graph is equal to:

$$\begin{aligned}\Delta f &= f(x + \Delta x) - f(x)\\ &= f(2+2) - f(2)\\ &= f(4) - f(2)\\ &= 16 - 4\\ &= 12\end{aligned}$$

Thus the average rate of change of the function over this interval is equal to

$$\frac{\Delta f}{\Delta x} = \frac{12}{2} = 6$$

This value is greater than the value we observed from $x = 0$ to $x = 2$. This implies that from $x = 2$ to $x = 4$, f(x) is increasing at a greater rate than from $x = 0$ to $x = 2$. This is despite the fact that in both cases the $\Delta x = 2$.

The rate of change is therefore not constant over the but is increasing with x.

When we say the rate of change of $f(x)$ from $x = 2$ to $x = 4$ is 6, it is only the average value for the given interval since it assumes rate of change is constant over that interval.

Let us move on to the next interval of $x = 4$ to $x = 6$. Once again the change in x or Δx equals 2 but the corresponding change in $f(x)$ is not the same as before.

$$\begin{aligned} \Delta f &= f(x + \Delta x) - f(x) \\ &= (x + \Delta x)^2 - (x)^2 \\ &= x^2 + 2x\Delta x + \Delta x^2 - x^2 \\ &= 2x\Delta x + \Delta x^2 \end{aligned}$$

For x equal to 4 and $\Delta x = 2$, the change in the function over this interval is

$$\begin{aligned} \Delta f &= 2x\Delta x + \Delta x^2 \\ \Delta f &= 2(4)(2) + (2)^2 \\ \Delta f &= 20 \end{aligned}$$

Note that this corresponds to the same value we would get by:

$$\begin{aligned} \Delta f &= f(6) - f(4) \\ &= 6^2 - 4^2 \\ &= 36 - 16 \\ &= 20 \end{aligned}$$

The average rate of change over this interval is therefore:

$$\frac{\Delta f}{\Delta x} = \frac{20}{2} = 10$$

This is still larger than the rate of change for the previous interval which was 6. Remember rate of change , by definition, refers to how much the function changes with respect to a change in the independent variable. The steepness or slope of the line over the interval provides a geometric understanding for this concept.

The average rate of changes calculated over each interval can be used to approximate the graph of *f*(*x*) from $x = 0$ *to* $x = 8$.

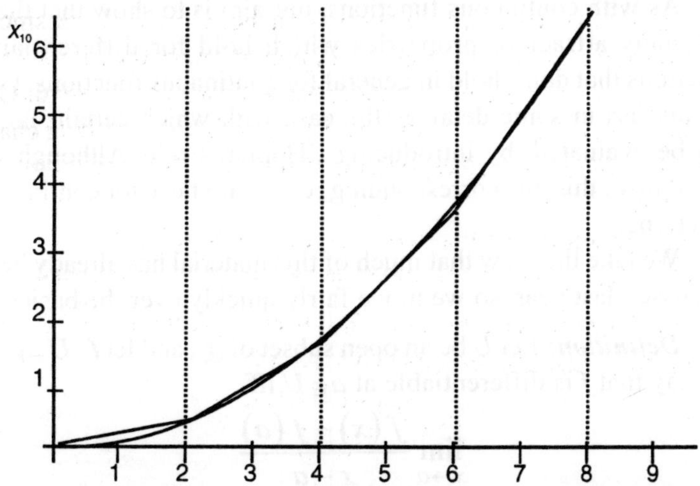

This roughly corresponds to the original graph, but as we can see the rate of change or slope is not constant through out the interval from $x = 0$ *to* $x = 8$, but increases as x-increases. In order to get more accurate answers we need to reduce our interval of $\Delta x = 2$ to a much smaller one. The idea being we need to analyze our graph over a small interval, Δx, to see what exactly is going on at each instant the function is changing. Here is where we begin our study of Calculus. We break down and freeze a changing situation into an infinite series of actions and analyzing what is going on in each individual actions.

BASIC PROPERTIES OF DERIVATIVE

We continue our study of classes of functions which are suitably restricted. Again we are passing from the general to the particular. The next most particular class of function we study after the class of continuous functions is the class of differentiable functions. We discuss the definition, show how to get "new functions from old" in what by now is a fairly routine way, and prove that this is a smaller class: that every differentiable function is continuous, but that there are continuous functions that are not differentiable. Informally, the difference is that the graph of a differentiable function may not have any sharp corners in it.

As with continuous functions, our aim is to show that there are many attractive properties which hold for differentiable functions that don't hold in general for continuous functions. One we discuss in some detail is the ease with which certain limits can be evaluated, by introducing l'Hopital's rule. Although we don't prove this, the corresponding results are false for continuous functions.

We take the view that much of this material has already been discussed last year, so we move fairly quickly over the basics.

Definition: Let U be an open subset of $\mathbb{R}$, and let $f: U \to \mathbb{R}$. We say that f is differentiable at $a \in U$ iff

$$\lim_{x \to \alpha} \frac{f(x) - f(a)}{x - a}$$

or equivalently,

$$\lim_{h \to 0} \frac{f(a+h) - f(a)}{h}$$

exists.

The limit, if it exists, is written as $f'(a)$.

We say that f is differentiable in U if and only if it is differentiable at each a in U.

Note that the Newton quotient is not defined when $x = a$, nor need it be for the definition to make sense. But the Newton quotient, if it exists, can be extended to be a continuous function at a by defining its value at that point to be $f'(a)$. Note also the emphasis on the existence of the limit. Differentiation is as much about showing the *existence* of the derivative, as *calculating* the value of the derivative.

Example: Let $f(x) = x^3$. Show, directly from the definition, that $f'(a) = 3a^2$.

Solution: This is just another way of asking about particular limits, we must compute

$$\lim_{x\to\alpha} \frac{x^3 - a^3}{x - a} = \lim_{x\to\alpha} \frac{(x-a)\left(x^2 + xa + a^2\right)}{x-a} = \lim_{x\to\alpha} (x^2 + 2xa + a^2) = 3a^2.$$

Exercise: Let $f(x) = \sqrt{x}$. Show, directly from the definition, that $f'(a) = 1/2\ \sqrt{a}$ when $a \neq 0$. What function do you have to consider in the particular case when $a = 4$?

Just as with continuity, it is impractical to use this definition every time to compute derivatives; we need results showing how to differentiate the usual class of functions, and we assume these are known from last year. Thus we assume the rules for differentiation of sums products and compositions of functions, together with the known derivatives of elementary functions such as sin, cos and tan; their reciprocals sec, *cosec* and cot; and exp and log.

Proposition: Let f and g be differentiable at a, and let k be a constant. Then $k.f$, $f + g$ and fg are differentiable at f. Also, if $g(a) \neq 0$, then f/g is differentiable at a. Let f be differentiable at a, and let g be differentiable at $f(a)$. Then gof is differentiable at a. In addition, the usual rules for calculating these derivatives apply.

Example: Let $f(x) = \tan\left(\dfrac{x^2 - a^2}{x^2 + a^2}\right)$ for $a \neq 0$. Show that f is differentiable at every point of its domain, and calculate the derivative at each such point.

Solution: This is the same example we considered in. There we showed the domain was the whole of $\mathbb{R}$, and that the function was continuous everywhere. Let $g(x) = \dfrac{x^2 - a^2}{x^2 + a^2}$. Then g is properly defined for all values of x, and the quotient is differentiable, since each term is, and since $x^2 + a^2 \neq 0$ for any x since $a \neq 0$. Thus f is differentiable using the chain rule since f= tanog, and we are assuming known that the elementary functions like tan are differentiable.

Finally to actually calculate the derivative, we have:

$$f'(x) = \sec^2\left(\frac{x^2 - a^2}{x^2 + a^2}\right)\frac{\left(x^2 + a^2\right)\cdot 2x - \left(\left(x^2 - a^2\right)\cdot 2x\right)}{\left(x^2 + a^2\right)^2}.$$

$$= \frac{4a^2x}{\left(x^2 + a^2\right)^2}\cdot \sec^2\left(\frac{x^2 - a^2}{x^2 + a^2}\right)$$

Exercise: Let $f(x) = \exp\left(\dfrac{1 + x^2}{1 - x^2}\right)$. Show that f is differentiable at every point of its domain, and calculate the derivative at each such point. The first point in our study of differentiable functions is that it is *more* restrictive for a function to be differentiable, than simply to be continuous.

Proposition: Let f be differentiable at a. Then f is continuous at a.

***Proof*:** To establish continuity, we must prove that $\lim_{x \to a} f(x) = f(a)$. Since the Newton quotient is known to converge, we have for $x \neq a$,

$$f(x) - f(a) = \frac{f(x) - f(a)}{x - a}.(x - a) \to f'(a).0 \quad \text{as } x \to a.$$

Hence f is continuous at a.

Example: Let $f(x) = |x|$; then f is continuous everywhere, but *not* differentiable at 0.

Solution: We already know from Example that $|x|$ is continuous. We compute the Newton quotient directly; recall that $|x| = x$ if $x \geq 0$, while $|x| = -x$ if $x < 0$. Thus

$$\lim_{x\to 0+} \frac{f(x)-f(0)}{x-0} = \lim_{x\to 0+} \frac{x-0}{x-0}, \text{ while}$$

$$\lim_{x\to 0-} \frac{f(x)-f(0)}{x-0} = \lim_{x\to 0-} \frac{-x-0}{x-0} = -1.$$

Thus both of the one-sided limits exist, but are unequal, so the limit of the Newton quotient does not exist.

Simple Limits

Our calculus of differentiable functions can be used to compute limits which otherwise prove troublesome.

Proposition: Let f and g be functions such that $f(a) = g(a) = 0$ while $f'(a)$ and $g'(a)$ both exits and $g'(a) \neq 0$. Then

$$\lim_{x\to a} \frac{f(x)}{g(x)} = \frac{f'(a)}{g'(a)}.$$

Proof: Since $f(a) = g(a) = 0$, provided $x \neq a$, we have

$$\frac{f(x)}{g(x)} = \frac{f(x)-f(a)}{g(x)-g(a)} = \frac{f(x)-f(a)}{x-a} \frac{x-a}{g(x)-g(a)}$$

$$\to \frac{f'(a)}{g'(a)} \text{ as } x \to a$$

where the last limit exists, since $g'(a) \neq 0$.

Remark: If $f'(a)$ and $g'(a)$ exist, $\lim_{x\to\alpha} \frac{f(x)}{g(x)}$ computing is easy, since f and g must be continuous at a by Proposition, unless we get an indeterminate form 0/0 or ∞/∞ for the formal quotient.

In fact l'Hopitals rule helps in both cases, although we need to develop stronger forms.

Example: Show that $\lim_{x \to 0} \frac{3x - \sin x}{x} = 2.$

Solution: Note first that we cannot get the result trivially from, since since $g(a) = 0$ and so we get the indeterminate form 0/0. However, we are in a position to apply the simple form of l'Hopital, since $x' = 1 \neq 0$. Applying the rule gives

$$\lim_{x \to 0} \frac{3x - \sin x}{x} = \lim_{x \to 0} \frac{3 - \cos x}{1} = 2.$$

Example: Show that $\lim_{x \to 0} \frac{\sqrt{1+x} - 1}{x} = 1/2.$

Solution: Note again that we cannot get the result trivially from 4.16, since this gives the indeterminate 0/0 form, because $g(a) = 0$. However, we are in a position to apply the simple form of l'Hopital, since $x' = 1 \neq 0$. Applying the rule gives

$$\lim_{x \to 0} \frac{\sqrt{1+x} - 1}{x} = \lim_{x \to 0} \frac{2^{-1}(1+x)^{-1/2}}{1} = 1/2.$$

Example: Show that $\lim_{x \to 0} \frac{\sin x}{x} = 1.$

Solution: This is spurious because we need the limit to calculate the derivative in the first place, but applying l'Hopital certainly gives the result.

Rolle and the Mean Value Theorem

We can combine our definition of derivative with the Intermediate Value Theorem to give a useful result which is in fact the basis of most elementary applications of the differential calculus. Like the results on continuous functions, it is a global result, and so needs continuity and differentiability on a whole interval.

Theorem (Rolle's Theorem): Let f be continuous on $[a, b]$, and differentiable on (a, b), and suppose that $f(a) = f(b)$. Then there is some c with $a < c < b$ such that $f'(c) = 0$.

Note: The theorem guarantees that the point c exists somewhere. It gives *no* indication of how to find c. Here is the diagram to make the point geometrically:

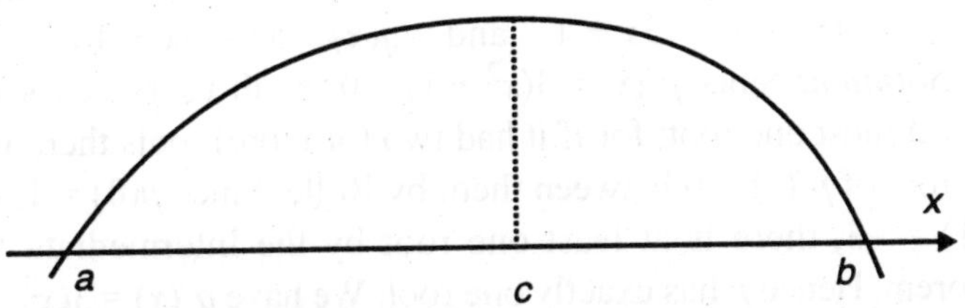

Fig. If f Crosses the Axis twice, somewhere between the two Crossings, the Function is flat. The Accurate Statement of this "Obvious" Observation is Rolle's Theorem.

***Proof*:** Since f is continuous on the compact interval $[a, b]$, it has both a global maximum and a global minimum. Assume first that the global maximum occurs at an interior point $c \in (a, b)$. In what follows, we pick h small enough so that $c + h$ always lies in (a, b). Then

If $h > 0$, $\dfrac{f(c+h)-f(c)}{h} \leq 0$, and so

$$\lim_{h\to 0+} \frac{f(c+h)-f(c)}{h} \leq 0,$$

since we know the limit exists.

Similarly, if $h < 0$, $\dfrac{f(c+h)-f(c)}{h} \geq 0$,

and so

$$\lim_{h\to 0+} \frac{f(c+h)-f(c)}{h} \geq 0,$$

since we know the limit exists. Combining these, we see that $f'(c) = 0$, and we have the result in this case. A similar argument applies if, instead, the global minimum occurs at the interior point c. The remaining situation occurs if both the global maximum and global minimum occur at end points; since $f(a) = f(b)$, it follows that f is constant, and any $c \in (a, b)$ will do.

Example: Investigate the number of roots of each of the polynomials

$$p(x) = x^3 + 3x + 1 \quad \text{and} \quad q(x) = x^3 - 3x + 1.$$

Solution: Since $p'(x) = 3(x^2 + 1) > 0$ for all $x \in \mathbb{R}$, we see that p has at most one root; for if it had two (or more) roots there would be a root of $p'(x) = 0$ between them by Rolle. Since $p(0) = 1$, while $p(-1) = -3$, there is at least one root by the Intermediate Value Theorem. Hence p has exactly one root. We have $q'(x) = 3(x^2 - 1) = 0$ when $x = \pm 1$. Since $q(-1) = 3$ and $q(1) = -1$, there is a root of q between -1 and 1 by the Intermediate Value Theorem. Looking as $x \to \infty$ and as $x \to -\infty$ shows here are three roots of q.

The Mean Value Theorem

Theorem: Let f be continuous on $[a, b]$, and differentiable on (a, b). Then there is some c with $a < c < b$ such that

$$\frac{f(b) - f(a)}{b - a} = f'(c)$$

or equivalently $f(b) = f(a) + (b - a)f'(c)$.

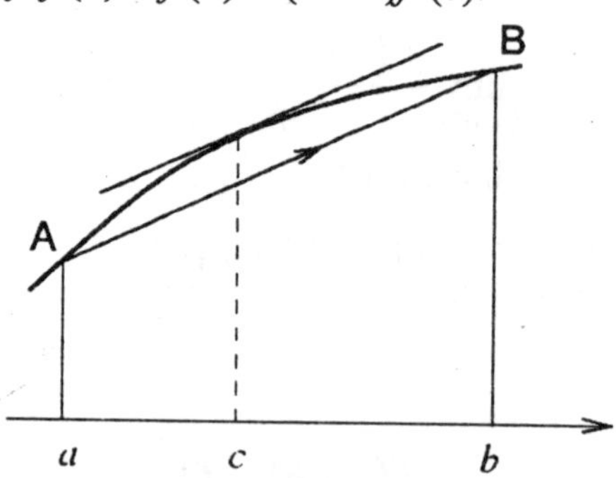

Fig. Somewhere Inside a chord, the Tangent to f will be Parallel to the chord.

The Accurate Statement of this Common-sense Observation is the Mean Value Theorem.

Proof: We apply Rolle to a suitable function; let

$$h(x) = f(b) - f(x) - \frac{f(b) - f(a)}{b - a}(b - x).$$

Then h is continuous on the interval $[a, b]$, since f is, and in the same way, it is differentiable on the open interval (a, b).

Also, $f(b) = 0$ and $f(a) = 0$. We can thus apply Rolle's theorem to h to deduce there is some point c with $a < c < b$ such that $h'(c) = 0$. Thus we have

$$0 = h'(c) = -f'(c) + \frac{f(b) - f(a)}{b - a},$$

which is the required result.

Example: The function f satisfies $f'(x) = \dfrac{1}{5 - x^2}$ and $f(0) = 2$. Use the Mean Value theorem to estimate f.

Solution: We first estimate the given derivative for values of x satisfying $0 < x < 1$. Since for such x, we have $0 < x^2 < 1$, and so $4 < 5 - x^2 < 5$. Inverting we see that

$$\frac{1}{5} < f'(x) < \frac{1}{4} \text{ when } 0 < x < 1.$$

Now apply the Mean Value theorem to f on the interval $[0, 1]$ to obtain some c with $0 < c < 1$ such that $f(1) - f(0) = f'(c)$. From the given value of $f(0)$, we see that $2.2 < f(1) < 2.25$

Exercise: The function f satisfies $f'(x) \dfrac{1}{5 + \sin x} =$ and $f(0) = 0$. Use the Mean Value theorem to estimate $f(\pi/2)$.

Note the "common sense" description of what we have done. If the derivative doesn't change much, the function will behave linearly. Note also that this gives meaning to the approximation

$$f(a + h) \approx f(a) + hf'(a).$$

We now see that the accurate version of this replaces $f'(a)$ by $f'(c)$ for some c between a and $a + h$.

The Cauchy Mean Value Theorem

Theorem: Let f and g be both continuous on $[a, b]$ and differentiable on (a, b). Then there is some point c with $a < c < b$ such that

$$g'(c)(f(b) - f(a)) = f'(c)(g(b) - g(a)).$$

In particular, whenever the quotients make sense, we have

$$\frac{f(b)-f(a)}{g(b)-g(a)} = \frac{f'(c)}{g'(c)}$$

Proof. Let $h(x) = f(x)(g(b) - g(a)) - g(x)(f(b) - f(a))$, and apply Rolle's theorem exactly as we did for the Mean Value Theorem. Note first that since both f and g are continuous on $[a, b]$, and differentiable on (a, b), it follows that h has these properties.

Also $h(a) = f(a)g(b) - g(a)f(b) = h(b)$. Thus we may apply Rolle to h, to deduce there is some point c with $a < c < b$ such that $h'(c) = 0$. But

$$h'(c) = f'(c)(g(b) - g(a)) - g'(c)(f(b) - f(a))$$

Thus

$$f'(c)(g(b) - g(a)) = g'(c)(f(b) - f(a))$$

This is one form of the Cauchy Mean Value Theorem for f and g. If $g'(c) \neq 0$ for any possible c, then the Mean Value theorem shows that $g(b) - g(a) \neq 0$, and so we can divide the above result to get

$$\frac{f(b)-f(a)}{g(b)-g(a)} = \frac{f'(c)}{g'(c)},$$

giving a second form of the result.

Note:Taking $g(x) = x$ recovers the Mean Value Theorem.

l'Hopital Revisited

We can get a much more useful form of l'Hopital's rule using the Cauchy Mean Value Theorem, directly from the definition of the derivative.

l'Hopital's Rule: General form

Proposition: Let f and g be functions such that $f(a) = g(a) = 0$, and suppose that f and g are differentiable on an open interval I containing a, and that $g'(x) \neq 0$, except perhaps at a. Then

$$\lim_{x\to\alpha} \frac{f(x)}{g(x)} = \lim_{x\to\alpha} \frac{f'(x)}{g'(x)},$$

provided the second limit exists.

Proof. Pick $x > a$ and apply the Cauchy Mean Value Theorem to the interval $[a, x]$, to find c with $a < c < x$ such that

$$\frac{f(x)}{g(x)} = \frac{f(x)-f(a)}{g(x)-g(a)}$$

$$= \frac{f'(c)}{g'(c)}$$

Then $\lim_{x\to a+} \frac{f(x)}{g(x)} = \lim_{c\to a+} \frac{f'(c)}{g'(c)} = \lim_{x\to a+} \frac{f'(x)}{g'(x)}$, since we know the actual limit (not just the one sided limit) exists. Now repeat with $x < a$ to get the result.

Example: Evaluate $\lim_{x\to 0} \frac{1-\cos x}{x^2} = \frac{1}{2}$.

Solution: We have

$$\lim_{x\to 0} \frac{1-\cos x}{x^2} = \lim_{x\to 0} \frac{\sin x}{2x} = \frac{1}{2}$$

where the use of l'Hopital is justified since the second limit exists. Note that you can't differentiate top and bottom again, and still expect to get the correct answer; one of the hypotheses of l'Hopital is that the first quotient is of the 0/0 form.

Example: Use l'Hopital to establish the following:

$$\lim_{x\to 0} \frac{\sqrt{1+x}-1-x/2}{x^2} = \frac{1}{8}$$

Solution: We have

$$\lim_{x\to 0}\frac{\sqrt{1+x}-1-x/2}{x^2}=\lim_{x\to 0}\frac{(1/2)(1+x)^{-1/2}-1/2}{2x}$$

$$=\lim_{x\to 0}\frac{(1/2)(-1/2)(1+x)^{-3/2}}{2}=\frac{-1}{8}$$

The use of l'Hopital is justified the second time, since the third limit exists; since we now know the second limit exists, the use of l'Hopital is justified the first time.

Infinite Limits

We can use Proposition to get results about infinite limits.

Example: Evaluate $\lim\limits_{x\to\infty} x\log\left(1+\frac{1}{x}\right)$

Solution: $\lim\limits_{x\to\infty} x\log\left(1+\frac{1}{x}\right) = \lim\limits_{x\to\infty}\frac{\log(1+y)}{y}$ writing $y = 1/x$,

$$=\lim_{y\to 0}\frac{\log(1+y)}{y} = \lim_{y\to 0}\frac{\log(1+y)}{y}=1$$

The last step is valid, since the final limit exists by l'Hopital; note also that this gives another way of finding $a_n = (1 + 1/n)^n$.

l'Hopital's Rule: Infinite Limits

Proposition: Let f and g be functions such that $\lim\limits_{x\to\infty} f(x) = \lim\limits_{x\to\infty} g(x) = \infty$, and suppose that $\lim\limits_{x\to\infty}\frac{f'(x)}{g'(x)}$ exists. Then

$$\lim_{x\to\infty}\frac{f(x)}{g(x)} = \lim_{x\to\infty}\frac{f'(x)}{g'(x)}$$

Proof: Pick $\in > 0$ and choose a such that

$$\lim_{x\to\infty}\left|\frac{f'(x)}{g'(x)}\right| - l < \in \quad \text{for all } x > a.$$

Then pick K such that if $x > K$, then $g(x) - g(a) \neq 0$. By Cauchy,

$$\frac{f'(c)}{g'(c)} = \frac{f(x)-f(a)}{g(x)-g(a)} \quad \text{for all } x > K.$$

Note that although c depends on x, we always have $c > a$. Then

$$\frac{f(x)}{g(x)} = \frac{f(x)-f(a)}{g(x)-g(a)} \cdot \frac{f(x)}{f(x)-f(a)} \cdot \frac{g(x)-g(a)}{g(x)}$$

TAYLOR'S THEOREM

We have so far explored the Mean Value theorem, which can be rewritten as

$$f(a + h) = f(a) + hf'(c)$$

where c is some point between a and $a + h$. We have already met the approximation

$$f(a + h) \sim f(a) + hf'(a)$$

when we studied the Newton - Raphson method for solving an equation, and have already observed that the Mean Value Theorem provides a more accurate version of this. Now consider what happens when f is a polynomial of degree n

$$f(x) = a_0 + a_1x + a_2x^2 + ... + a_{n-1}x^{n-1} + a_nx^n.$$

Note that $f(0) = a_0$. Differentiating gives

$$f'(x) = a_1 + 2a_2x + 3a_3x^2 ... + (n-1)a_{n-1}x^{n-2} + na_nx^{n-1},$$

and so $f'(0) = a_1$. Again, we have

$$f''(x) = 2a_2 + 3.2a_3x ... + (n-1)(n-2)a_{n-1}x^{n-3}$$
$$+ n(n-1)a_nx^{n-2},$$

and $f''(0) = 2a_2$. After the next differentiation, we get $f'''(0) = 3!a_3$, while after k differentiations, we get, $f^{(k)}(0) = k!a_k$, provided $k \leq n$. Thus we can rewrite the polynomial, using its value, and the value of its derivatives at 0, as

$$f(x) = f(0) + f'(0)x + \frac{f''(0)}{2!}x^2 + \frac{f'''(0)}{3!}$$

$$x^3 + ... + \frac{f^{(n-1)}(0)}{(n-1)!}x^{n-1} + \frac{f^{(n)}(0)}{n!}x^n.$$

This opens up the possibility of representing more general functions than polynomials in this way, and so getting a generalisation of the Mean Value Theorem.

Taylors Theorem - Lagrange form of Remainder

Theorem: Let f be continuous on $[a, x]$, and assume that each of $f', f'', ..., f^{(n+1)}$ is defined on $[a, x]$. Then we can write

$$f(x) = P_n(x) + R_n(x),$$

where $P_n(x)$, the Taylor polynomial of degree n about a, and $R_n(x)$, the corresponding remainder, are given by

$$P_n(x) = f(a) + f'(a)(x-a) + \frac{f''(a)}{2!}(x-a)^2$$

$$+ \frac{f^{(n)}(a)}{n!}(x-a)^n,$$

$$R_n(x) = \frac{f^{(n+1)}(c)}{(n+1)!}(x-a)^{n+1},$$

where c is some point between a and x. We make no attempt to prove this, although the proof can be done with the tools we have at our disposal. Some quick comments:

1. The theorem is also true for $x < a$; just restate it for the interval $[x, a]$ etc;
2. If $n = 0$, we have $f(x) = f(a) + (x - a)f'(c)$ for some c between a and x; this is a restatement of the Mean Value Theorem;

3. If $n = 1$, we have

$$f(x) = f(a) + (x - a)f'(x) + \frac{f''(c)}{2!}(x - a)^2$$

for some c between a and x; this often called the Second Mean Value Theorem;

4. In general we can restate Taylor's Theorem as

$$f(x) = f(a) + (x - a)f'(x) + ... + \frac{f^{(n)}(a)}{2!}(x - a)^n$$

$$+ \frac{f^{(n+1)}(c)}{(n+1)!}(x - a)^{n+1},$$

for some c between a and x;

5. The special case in which $a = 0$ has a special name; it is called Maclaurin's Theorem;

6. Just as with Rolle, or the Mean Value Theorem, there is no useful information about the point c.

We now explore the meaning and content of the theorem with a number of examples.

Example: Find the Taylor polynomial of order n about 0 for $f(x) = e^x$, and write down the corresponding remainder term.

Solution: There is no difficulty here in calculating derivatives clearly $f^{(k)}(x) = e^x$ for all k, and so $f^{(k)}(0) = 1$. Thus by Taylor's theorem,

$$e^x = 1 + x + \frac{x^2}{2!} + \frac{x^2}{2!} + ... \frac{x^n}{n!} + \frac{x^{n+1}}{(n+1)!}e^c$$

for some point c between 0 and x. In particular,

$$P_n(x) = 1 + x + \frac{x^2}{2!} + \frac{x^2}{2!} + ... \frac{x^n}{n!} \text{ and } R_n(x) =$$

$$R_n(x) = \frac{x^{n+1}}{(n+1)!}e^c.$$

We can actually say a little more about this example if we recall that x is *fixed.* We have

$$e^x = P_n(x) + R_n(x) = P_n(x) + \frac{x^{n+1}}{(n+1)!}e^c$$

We show that $R_n(x) \to 0$ as $n \to \infty$, so that (again for fixed x), the sequence $P_n(x) \to e^x$ as n. If $x < 0$, $e^c < 1$, while if $x \geq 1$, then since $c < x$, we have $e^c < e^x$. thus

$R_n(x) = e\max(\ e,\ 1)0$ as n.

$$\left|R_{n(x)}\right| = \left|\frac{x^{n+1}}{(n+1)!}e^c\right| \leq \frac{|x|^{n+1}}{(n+1)!}$$

max $(e^x, 1) \to 0$ as $n \to \infty$.

We think of the limit of the polynomial as forming a series, the Taylor series for e^x. We study series (and then Taylor series).

Example: Find the Taylor polynomial of order 1 about a for $f(x) = e^x$, and write down the corresponding remainder term.

Solution: Using the derivatives computed above, by Taylor's theorem,

$$e^x = e^\alpha + (x-a)e^\alpha + \frac{(x-a)^2}{2!}e^c$$

for some point c between a and x. In particular,

$$P_1(x) = e^\alpha + (x-a)e^\alpha \text{ and } R_1(x) = \frac{(x-a)^2}{2!}e^c.$$

Example: Find the Maclaurin polynomial of order $n > 3$ about 0 for $f(x) = (1+x)^3$, and write down the corresponding remainder term.

Solution: We have

$$f(x) = (1+x)^3$$
$$f'(x) = 3(1+x)^2$$
$$f''(x) = 6(1+x)$$

$$f'''(x) = 6$$
$$f^{(n)}(x) = 0 \text{ if } n > 3.$$
$$f(0) = 1$$
$$f'(x) = 3$$
$$f''(x) = 6$$
$$f'''(x) = 6$$

and so, by Taylor's theorem

$$(1 + x)^3 = 1 + 3x + \frac{6}{2!}x^2 + \frac{6}{3!}x^3,$$

a result we could have got directly, but which is at least reassuring.

Find the Taylor polynomial of order n about 0 for $f(x) = \sin x$, and write down the corresponding remainder term.

Solution: There is no difficulty here in calculating derivatives we have

$$f(x) = \sin x$$
$$f'(x) = \cos x$$
$$f''(x) = -\sin x$$
$$f'''(x) = -\cos x$$
$$f^{(4)}(x) = \sin x \text{ and so on.}$$
$$f(0) = 0$$
$$f'(x) = 1$$
$$f''(x) = 0$$
$$f'''(x) = -1$$

Thus by Taylor's theorem,

$$\sin x = x - \frac{x^3}{3!} + \frac{x^5}{5!} + \ldots + (-1)^{n+1}\frac{x^{2n+1}}{(2n+1)!}$$

Writing down the remainder term isn't particularly useful, but the important point is that

$$|R_{2n+1}(x)|\left|\frac{x^{2n+3}}{(2n+3)!}\right| \to 0 \text{ as } n \to \infty.$$

Exercise: Recall that $\cosh x = \dfrac{e^x + e^{-x}}{2}$, and that sinh,

$$x = \frac{e^x + e^{-x}}{2}.$$

Now check the shape of the following Taylor polynomials:

$$\cos x = 1 - \frac{x^2}{2!} + \frac{x^4}{4!} + \ldots + (-1)^n \frac{x^{2n}}{2n!} + \ldots$$

$$\sinh x = x + \frac{x^3}{3!} + \frac{x^5}{5!} + \ldots + \frac{x^{2n+1}}{(2n+1)!} + \ldots$$

$$\cosh x = 1 + \frac{x^2}{2!} + \frac{x^4}{4!} + \ldots + \frac{x^{2n}}{2n!} + \ldots$$

Example: Find the maximum error in the approximation

$$\sin(x) \sim x - \frac{x^3}{3!} \text{ given that } |x| < 1/2.$$

Solution: We use the Taylor polynomial for sin x of order 4 about 0, together with the corresponding remainder. Thus

$$\sin x = x - \frac{x^3}{3!} + \frac{x^5}{5!} \cos c$$

for some c with $0 < c < 1/2$ or $-1/2 < c < 0$. In any case, since $|x| < 1/2$,

$$\left|\frac{x^5}{5!} \cos c\right| \le \left|\frac{x^5}{5!}\right| \le \left|\frac{1}{2^5.5!}\right| \le \frac{1}{120.32}$$

Warning: The Taylor polynomial always exists, providing f is suitably differentiable. But it need not be useful. Consider the example

$$f(x) = \begin{cases} \exp\left(-1/x^2\right) \text{if } x > 0; \\ 0 \quad\quad \text{if } x \le 0 \end{cases}$$

The interest in f is at 0; it is well behaved everywhere else. It turns out that,

$$f(0) = f'(0) = f'(0) = ... = f^{(n)}(0) = ... = 0.$$

So the Taylor polynomial of degree n for f about 0 is $P_n(x) = 0 + 0x + 0x^2 + ... + 0.x^n = 0$, and so for every n, $R_n(x) = f(x)$. Clearly in this case, P_n tells us nothing useful about the function.

Example: Find the Taylor polynomial of order n about 0 for $f(x) = (1 + x)^\alpha$, and note that this gives a derivation of the binomial theorem. In fact, the remainder $|R_n(x)| \to 0$ as $n \to \infty$, provided $|x| < 1$.

Solution: There is again no difficulty here in calculating derivatives we have

$$f(x) = (1+x)^\alpha$$

$$f'(x) = \alpha(1+x)^{\alpha-1}$$

$$f''(x) = \alpha(\alpha-1)(1+x)^{\alpha-2}$$

$$f'''(x) = \alpha(\alpha-1)(\alpha-2)(1+x)^{\alpha-3}$$

and in general

$$f^{(n)}(x) = \alpha(\alpha-1)...(\alpha-n+1)(1+x)^{\alpha-n}$$

$$f(0) = 1$$

$$f'(x) = \alpha$$

$$f''(x) = \alpha(\alpha-1)$$

$$f'''(x) = \alpha(\alpha-1)(\alpha-2)$$

$$f^{(n)} = \alpha(\alpha-1)...(\alpha-n+1)$$

$$f(x) = (1+x)^\alpha$$

$$f'(x) = \alpha(1+x)^{\alpha-1}$$

$$f''(x) = \alpha(\alpha-1)(1+x)^{\alpha-2}$$

$$f'''(x) = \alpha(\alpha-1)(\alpha-2)(1+x)^{\alpha-3}$$

and in General,

$$f^{(n)}(x) = \alpha(\alpha-1)\ldots(\alpha-n+1)(1+x)^{\alpha-n}$$
$$f(0) = 1$$
$$f'(x) = \alpha$$
$$f''(x) = \alpha(\alpha-1)$$
$$f'''(x) = \alpha(\alpha-1)(\alpha-2)$$
$$f^{(n)} = \alpha(\alpha-1)\ldots(\alpha-n+1)$$

Thus by Taylor's theorem,

$$(1+x)^{\alpha} = 1 + \alpha x - \frac{\alpha(\alpha-1)}{2!}x^2 + \frac{\alpha(\alpha-1)(\alpha-2)}{3!}x^3 +$$
$$\ldots + \frac{\alpha(\alpha-1)\ldots(\alpha-n+1)}{n!}x^n + \ldots$$

The remainder is not hard to deal with, but we omit the proof; in fact $|R_n(x) \to 0|$ when $n \to \infty$.

Note also that if $\alpha > 0$ is an *integer*, say $\alpha = n$ then $|R_n(x)| = 0$ and $f(x) = P_n(x)$. This is another way to get the Binomial theorem described.

INFINITE SERIES

In this section we return to study a particular kind of sequence, those built by adding up more and more from a given collection of terms. One motivation comes from section, in which we obtained a sequence of approximating polynomials $\{P_n\}$ to a given function. It is more natural to think of adding additional terms to the polynomial, and as such we are studying series. However, there is a closely related sequence; the sequence of partial sums.

Arithmetic and Geometric Series

Consider the Sum

$$a + (a + r) + (a + 2r) + \cdots + (a + nr)$$

We are trying to add up the terms in an arithmetic progression. A small amount of notation makes the addition easy.

Let $S_n = a + (a + r) + (a + 2r) + \cdots + (a + nr)$,

$S_n = (a + nr) + (a + (n - 1)r) + \cdots + (a + r) + a$,

so $2S_n = (2a + nr) + (2a + nr) + \cdots + (2a + nr)$, and

$$S_n = (n + 1).\frac{(\alpha + (\alpha + nr))}{2}.$$

Note that if $r > 0$ then $S_n \to \infty$ as $n \to \infty$, while if $r < 0$ then $S_n \to -\infty$ as $n \to \infty$.

We next consider a geometric progression (or series):

Let $S_n = a + ar + ar^2 + \cdots + ar^n$,

$rS_n = ar + ar^2 + ar^3 + \cdots + ar^{n+1}$,

so $(1 - r) S_n = a(1 - r^{n+1})$, and

$$S_n = \frac{a(1-r^{n+1})}{1-r} \quad \text{if} \quad r \neq 1.$$

Note that if $|r| < 1$, then $S_n \to \dfrac{a}{1-r}$ as $n \to \infty$.

If $r > 1$, say $r = 1 + k$,

$$S_n = a\left(\frac{(1+k)^{n+1} - 1}{k}\right),$$

$$> \frac{a}{k}\left(1 + (n+1)_{k-1}\right) > a(n+1) \to \infty \text{ if } a > 0.$$

Convergent Series

Definition: Let $\{a_n\}$ be a sequence, and let,

$S_n = a_1 + a_2 + \cdots + a_n$, the nth partial sum.

If $\lim_{n\to\infty} s_n$ exists, we say that $\sum a_n$ is a convergent series. and write $\lim_{n\to\infty} s_n = \sum a_n$.

Thus a series is convergent if and only if it's sequence of partial sums is convergent. The limit of the sequence of partial sums is the sum of the series. A series which is *not* convergent, is a divergent series.

Example: The series $\sum r^n$ is convergent with sum $1/(1-r)$, provided that $|r|<1$. For other values of r, the series is divergent; in particular, the series $\sum(-1)^n$ is divergent.

Solution: We noted above that when $|r|<1$, $S_n \to a/(1-r)$ as $n \to \infty$; note particular cases;

$$\sum_{n=1}^{\infty}\frac{1}{2^n}=1 \text{ or equivalently, } \frac{1}{2}+\frac{1}{4}+\frac{1}{8}+\ldots = 1.$$

Example: The sum $\sum \frac{1}{n(n+1)}$ is convergent with sum 1.

Solution: We can compute the partial sums explicitly:

$$S_n=\sum_{k=1}^{n}\frac{1}{k(k+1)}=\sum_{k=1}^{n}\left(\frac{1}{k}-\frac{1}{k+1}\right)=1-\frac{1}{n+1}\to 1 \text{ as } n\to\infty$$

Example: The sum $\sum \frac{1}{n}$ is divergent.

Solution: We estimate the partial sums:

$$\begin{aligned}S_n&=\frac{1}{1}+\frac{1}{2}+\left(\frac{1}{3}+\frac{1}{4}\right)+\left(\frac{1}{5}+\ldots+\frac{1}{8}\right)+\left(\frac{1}{9}+\ldots+\frac{1}{15}\right)+\ldots+\frac{1}{n}\\&>1+\frac{1}{2}+\frac{2}{4}+\frac{4}{8}>2\frac{1}{2} \text{ if } n\geq 15\\&>1+\frac{1}{2}+\frac{2}{4}+\frac{4}{8}+\frac{8}{16}>3 \text{ if } n\geq 31\\&\to \infty \text{ as n} \to \infty .\end{aligned}$$

Example: The sum is $\sum \frac{1}{n^2}$ convergent. [Actually the sum is $\frac{\pi^2}{6}$, but this is much harder.]

Solution: We estimate the partial sums. Since $1/n^2 > 0$, clearly $\{S_n\}$ is an increasing sequence. We show it is bounded above, whence by the Monotone Convergence Theorem, it is convergent. From the diagram,

$$\frac{1}{2^2}+\frac{1}{3^2}+\ldots+\frac{1}{n^2} = \int_1^n \frac{dx}{x^2}, \text{ and so}$$

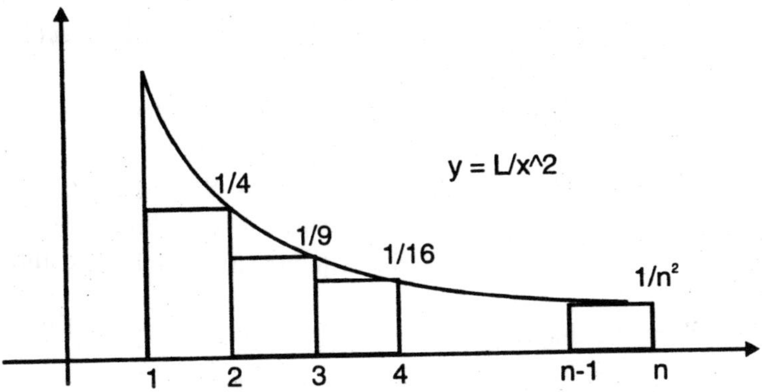

Fig. Comparing the area under the Curve $y = 1/x^2$ with the area of the Rectangles Below the Curve

$$S_n < 1 + \left[-\frac{1}{x}\right]_1^n \le 2 - \frac{1}{n}$$

Thus $S_n < 2$ for all n, the sequence of partial sums is bounded above, and the series is convergent.

Proposition: Let $\sum a_n$ be convergent. Then $a_n \to 0$ as $n \to \infty$.

Proof: Write $l = \lim_{n\to\infty} s_n$, and recall from our work on limits of sequences that $S_{n-1} l \to l$ as $n \to \infty$.

Then,

$$a_n = (a_1 + a_2 + \ldots a_n) - (a_1 + a_2 + \ldots a_{n-1})$$
$$= S_n - S_{n-1} l - l \quad \text{as} \quad n.$$

Example: The sum $\sum \frac{1}{n}$ is divergent (Graphical method).

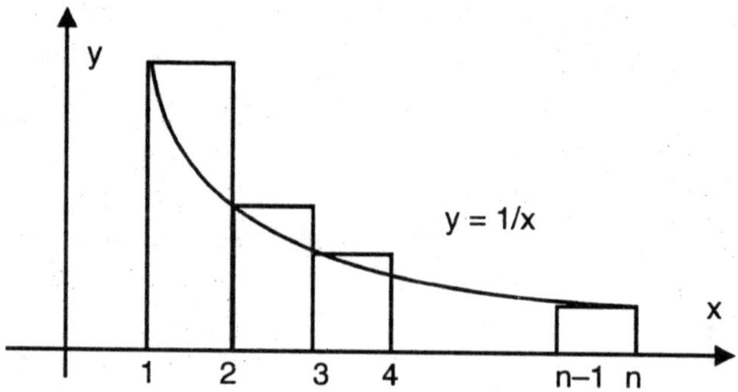

Fig. Comparing the area under the Curve $y = 1/x$ with the area of the Rectangles above the Curve

Solution: We estimate the partial sums. Since $1/n > 0$, clearly $\{S_n\}$ is an increasing sequence. We show it is not bounded above, whence by the note after 3.9, the sequence of partial sums $\to \infty$ as $n \to \infty$.

From the diagram,

$$1+\frac{1}{2}+\ldots+\frac{1}{n}>\int_1^n \frac{dx}{x}>\frac{1}{2}+\ldots+\frac{1}{n}$$

Writing $S_n = 1 + \frac{1}{2} + \cdots + \frac{1}{n}$, we have $S_n > \log n > S_n - 1$, or equivalently $1 + \log n > S_n > \log n$ for all n. Thus $S_n \to \infty$ and the series is divergent.

Proposition: Let $\sum a_n$ and $\sum b_n$ be convergent. Then $\sum (a_n + b_n)$ and $\sum c.a_n$ are convergent.

Proof: This can be checked easily directly from the definition; it is in effect the same proof that the sum of two convergent sequences is convergent etc

THE COMPARISON TEST

We have already used the Monotone Convergence Theorem in studying simple series. In fact it is a lot more useful. When we know the behaviour of some simple series, we can deduce many more results by comparison as follows.

Theorem: Assume that $0 \leq a_n \leq b_n$ for all n, and suppose that $\sum b_n$ is convergent. Then a_n is convergent.

Proof: Define

$$S_n = a_1 + a_2 + \cdots + a_n,$$
$$T_n = b_1 + b_2 + \cdots b_n.$$

Then by hypothesis, $0 \leq S_n \leq T_n$. Since $\{T_n\}$ is a convergent sequence, it is a bounded sequence.

In particular, it is bounded above, so there is some K such that $T_n \leq K$ for all n. Thus $S_n \leq K$ for all n, so $\{S_n\}$ is a sequence that is bounded above; since $a_n \geq 0$, $S_{n+1} = S_n + a_n \geq S_n$ and $\{S_n\}$ is an increasing sequence. Thus by the Monotone Convergence Theorem, it is a convergent sequence.

Example: Let $a_n = \dfrac{2n}{3n^3 - 1}$ and let $b_n = \dfrac{1}{n^2}$. Then $\sum a_n$ is convergent

For $n \geq 1$, $n^3 \geq 1$, so $3n^3 - 1 \geq 2n^3$. Thus

$$a_n = \frac{2n}{3n^3 - 1} \leq \frac{2n}{2n^3} = b_n$$

Since we know that $\sum b_n$ is convergent, so is $\sum a_n$.

Remark: The conclusions of *Theorem* remain true even if we only have $a_n \leq b_n$ eventually; for if it holds for $n \geq N$, we replace the inequality by

$$S_n \leq T_n + a_1 + a_2 + \cdots + a_N$$

and this then holds for all n.

Example: Let $a_n = \dfrac{\log n}{n}$, and compare with $b_n = \dfrac{1}{n}$.

Solution: Note that if $n \geq 3$, then $\log n > 1$. We can thus use the "eventually" form of the comparison test; we have

$$a_n = \frac{\log n}{n} > \frac{1}{n} = b_n,$$

We deduce divergence, for if $\sum a_n$ were convergent, it follows that that $\sum b_n$ was convergent.

Limiting form of the Comparison Test

Corollary: Suppose that $a_n > 0$ and $b_n > 0$, and that there is some constant k such that $\lim_{n\to\infty} \frac{a_n}{b_n} = k > 0$. Then $\sum a_n$ is convergent iff $\sum b_n$ is convergent.

Proof: Assume first that $\sum b_n$ is convergent. Since $a_n/b_n \to k$ as $n \to \infty$, eventually (take $\in = k > 0$), we have $a_n \le 2kb_n$, which is convergent. Hence $\sum a_n$ is convergent. To get the converse, note that $b_n/a_n \to 1/k$ as $n \to \infty$, so we can use the same argument with a_n and b_n interchanged.

Example: Let $a_n = \frac{n}{n^2+1}$ and let $b_n = \frac{1}{n}$. Then $\sum a_n$ is divergent by the limiting form of the comparison test.

Solution: Note that the terms are all positive, so we try to apply the limiting form of the comparison test directly.

$$\frac{a_n}{b_n} = \frac{n}{n^2+1} \cdot \frac{n}{1} = \frac{n^2}{n^2+1} \to 1 \text{ as } n \to \infty$$

Since the limit is non-zero, the use of the limiting form of the comparison test is valid, and we see that $\sum a_n$ is divergent.

This is all very well, but as with the "new sequences from old" programme, we need a few reference sequences before we can get further. One set is the geometric series, which we have already met.

Exercise: Let $a_n =$ and let $b_n = \left(\frac{2}{3}\right)^n$. Use the limiting form of the comparison test to show that $\sum a_n$ is convergent.

We also know about $\sum 1/n\alpha$, at least when $\alpha \geq 2$ when it converges, by comparison with $\sum 1/n^2$, and when $\alpha \leq 1$ when it diverges, by comparison with $\sum 1/n$.

Proposition: The sum $\sum \frac{1}{n^{\alpha}}$ is convergent when $\alpha > 1$.

Solution: Assume $\alpha > 1$; we estimate the partial sums. Since $1/n^{\alpha} > 0$, clearly $\{S_n\}$ is an increasing sequence.

Let $S_n = 1 + \frac{1}{2^{\alpha}} + \cdots + \frac{1}{n^{\alpha}}$, and consider the graph of $y = 1/x^{\alpha}$, noting that y is a decreasing function of x (which is where we use that fact that $\alpha > 1$). From a diagram which is essentially the same as that of Fig, shaded area $\int_1^n \frac{dx}{x^{\alpha}}$

$$\frac{1}{2^{\alpha}} + \ldots + \frac{1}{n^{\alpha}} < \left[\frac{1}{1-\alpha}\frac{1}{x^{\alpha-1}}\right]_1^n$$

$$S_n - 1 < \frac{1}{\alpha - 1}\left(1 - \frac{1}{n^{\alpha-1}}\right)$$

$$S_n < \frac{1}{\alpha - 1} + 1$$

Thus the sequence of partial sums is bounded above, and the series converges.

The Ratio

Theorem: Let $\sum a_n$ be a series, and assume that lim $\frac{|a_n + 1|}{|a_n|} \to r$ as $n \to \infty$. Then if $r < 1$, the series is convergent, if $r > 1$, the series is divergent, while if $r = 1$, the test gives no information.

Example: Let $a_n = \frac{2^n (n!)^2}{(2n)!}$. Then $\sum a_n$ is convergent.

Solution: We look the ratio of adjacent terms in the series (of positive terms).

$$\frac{(n+1)^{st}\text{ term}}{n^{th}\text{ term}} = \frac{a_{n+1}}{a_n} = \frac{2^{n+1}(n+1)!(n+1)!}{(2n+2)!}\,\frac{(2n)!}{2^n n!n!}$$

$$= \frac{2(n+1)^2}{(2n+2)(2n+1)} = \frac{n+1}{2n+1} \to \frac{1}{2} \text{ as } n \to \infty.$$

Since the ratio of adjacent terms in the series tends to a limit which is < 1, the series converges by the ratio test.

Absolute and Conditional Convergence

So far most of our work has been with series all of whose terms are positive. There is a good reason for this; there is very little we can say about series with mixed signs. Indeed there is just one useful result at this level, which is the topic of this section.

The easiest case occurs when the series really can be thought of as a series of positive terms.

Definition: The series $\sum_{n=1}^{\infty} a_n$ is absolutely convergent iff the series $\sum_{n=1}^{\infty} |a_n|$ is convergent.

Definition: The series $\sum_{n=1}^{\infty} a_n$ is conditionally convergent if and only if the series $\sum_{n=1}^{\infty} a_n$ is convergent but not absolutely convergent.

Example: Show that the series $\sum_{n=1}^{\infty} \frac{(-1)^{n+1}}{n^2}$ is absolutely convergent.

Solution: We have $\sum_{n=1}^{\infty} \left|\frac{(-1)^{n+1}}{n^2}\right| = \sum_{n=1}^{\infty} \frac{1}{n^2}$ and this is convergent. So $\sum_{n=1}^{\infty} \frac{(-1)^{n+1}}{n^2}$ is absolutely convergent.

Note: We choose to work with the sign $(-1)^{n+1}$ rather than $(-1)^n$ simply for tidiness; it is usual to start a series with a positive term, so the coefficient of a_1 is chosen to be +. Thus that of a_2 must be - etc if the series is to alternate.

Our interest in absolutely convergent series starts by observing that they are in fact all convergent. Indeed this is the easiest way to show a series is convergent if the terms are not all positive.

Proposition: An absolutely convergent series is convergent.

Proof: Assume that $\sum_{n=1}^{\infty} a_n$ is absolutely convergent, and define

$$a_n^+ = \begin{cases} a_n \text{ if } a_n > 0 \\ 0 \text{ if } a_n \le 0 \end{cases} \quad \text{and} \quad a_n^- = \begin{cases} |a_n| & \text{if } a_n < 0 \\ 0 & \text{if } a_n \ge 0 \end{cases}$$

The point of this definition is that

$$0 \le a_n + \le |a_n| \text{ and } 0 \le a_n - |a_n| \text{ for all n,}$$

so we have two new series of positive terms, while

$$|a_n| = a_n^+ + a_n^- \quad \text{and} \quad a_n = a_n^+ - a_n^-.$$

Using equation to compare with the convergent series $\sum_{n=1}^{\infty} |a_n|$, we see that each of

$$\sum_{n=1}^{\infty} a_n^+ \text{ and } \sum_{n=1}^{\infty} a_n^-$$

is a convergent series of positive terms. Thus

$$\sum_{n=1}^{\infty} a_n = \sum_{n=1}^{\infty} a_n + -a_n^- = \sum_{n=1}^{\infty} a_{n^+} + \sum_{n=1}^{\infty} a_n^-$$

is also convergent using.

This gives one way of proving that a series is convergent even if the terms are not all positive, and so we can't use the comparison test directly. There is essentially only one other way, which is a very special, but useful case known as Leibniz theorem, or the theorem on alternating signs, or the alternating series test. We give the proof because the argument is so like the proof of the convergence of the ratio of adjacent terms in the Fibonacci series.

Warning:Note how we usually talk about the "Fibonacci series", even though it is a *sequence* rather than a series. Try not to be confused by this popular but inaccurate usage.

Leibniz Theorem

Theorem: Let $\{a_n\}$ be a decreasing sequence of positive terms such that $a_n 0$ as $n \to \infty$. Then the series

$$\sum_{n=1}^{\infty}(-1)^{n+1} a_n \text{ is convergent.}$$

Proof: Write S_n for the nth partial sum of the series $\sum_{n=1}^{\infty}(-1)^{n+1} a_n$. We show this sequence has the same type of oscillating behaviour that the corresponding sequence of partial sums in the Fibonacci example. By definition, we have

$$s_{2n+1} = a_1 - a_2 + a_3 - ... + a_{2n-1} - a_{2n} + a_{2n+1}$$

$$s_{2n-1} = a_1 - a_2 + a_3 - ... + a_{2n-1}$$

and so, subtracting, we have

$$s_{2n+1} = s_{2n+1} - a_{2n} + a_{2n+1}.$$

Since $\{a_n\}$ is a decreasing sequence, $a_{2n} > a_{2n+1}$ and so $s_{2n+1} < s_{2n-1}$. Thus we have a decreasing sequence

$$s_1 > s_3 \quad > s_5 > ... > s_{2n-1} > s_{2n+1} > ....$$

Similarly $s_{2n} > s_{2n-2}$ and we have an increasing sequence

$$s_2 < s_4 < s_6 ... < s_{2n-2} < s_{2n} < ....$$

Also

$$s_{2n+1} = s_{2n} + a_{2n+1} > s_{2n}$$

Thus

$$s_2 < s_4 < s_6 ... < s_{2n-2} < s_{2n} < s_{2n+1} < s_{2n-1} < ... s_5 < s_3 < s_1$$

and the sequence $s_1, s_3, s_5, \ldots$ is a decreasing sequence which is bounded below (by s_2), and is convergent to α(say). Similarly $s_2, s_4, s_6, \ldots$ is an increasing sequence which is bounded above (by s_1), and s is convergent to β (say) also

$$s_{2n+1} - s_{2n} = a_{2n+1}$$

and so letting $n \to \infty$

$$\alpha - \beta = 0$$

So $\alpha = \beta$, and all the partial sums are tending to , so the series converges.

Example: Show that the series $\sum_{n=1}^{\infty} \frac{(-1)^{n+1}}{n}$ is conditionally convergent.

Solution: We have $\sum_{n=1}^{\infty} \left| \frac{(-1)^{n+1}}{n} \right| = \sum_{n=1}^{\infty} \frac{1}{n}$ and this is divergent; thus the series is *not* absolutely convergent.

We show that this series is still convergent, and so is conditionally convergent.

Write $a_n = 1/n$, so $a_n > 0$, $a_{n+1} < a_n$ and $a_n \to 0$ as $n \to \infty$. Thus all the conditions of Leibniz's theorem are satisfied, and so the series $\sum_{n=1}^{\infty} \frac{(-1)^{n+1}}{n}$ is convergent.

Re-arranging an Absolutely convergent Series

Proposition: Let $\sum_{n=1}^{\infty} a_n$ be an absolutely convergent series and suppose that $\{b_n\}$ is a re-arrangement of $\{a_n\}$. Then $\sum_{n=1}^{\infty} b_n$ is convergent, and

$$\sum_{n=1}^{\infty} b_n = \sum_{n=1}^{\infty} a_n .$$

Warning: It is not useful to re-arrange conditionally convergent series. There is a result which is an extreme form of this: Pick $x \in \mathbb{R}$, and let a_n be a conditionally convergent series. then there is a re-arrangement $\{b_n\}$ of $\{a_n\}$ such that $\sum_{n=1}^{\infty} b_n = x$! In other words, we can re-arrange to get any answer we want!

An Estimation Problem

This section shows how we can use a lot of the earlier ideas to produce what is often wanted in practice—results which are an approximation, together with an indication of how good the approximation is. Find how accurate the approximation obtained by just using the first ten terms is, to,

$$\sum_{n=1}^{\infty} \frac{1}{n^2}.$$

Again we are going to use geometrical methods for this. Our geometric statement follows from the diagram, and is the assertion that the area of the rectangles below the curve is less than the area under the curve, which is less than the area of the rectangles which contain the curve.

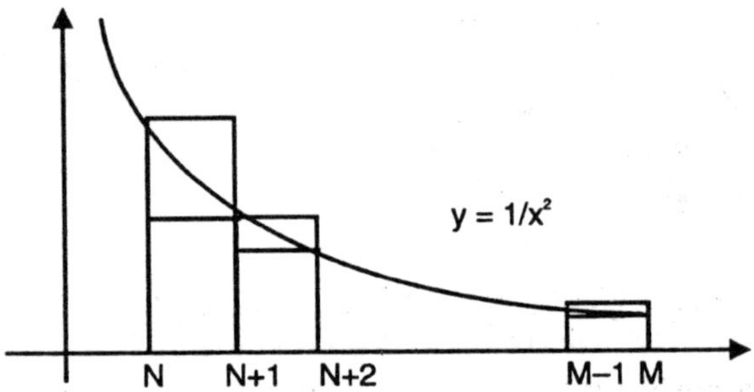

Fig. An upper and Lower Approximation to the area under the Curve

Writing this out geometrically gives the statement:

$$\sum_{n=N+1}^{\infty} \frac{1}{n^2} \le \int_N^M \frac{dx}{x^2} \le \sum_{n=N}^{M-1} \frac{1}{n^2}$$

We can evaluate the middle integral:

$$\int_N^M \frac{dx}{x^2} = \left[-\frac{1}{x}\right]_N^M = \frac{1}{N} - \frac{1}{M}$$

For convenience, we define

$$S = \sum_{n=1}^{\infty} \frac{1}{n^2} \text{ and } S_N = \sum_{n=1}^{N} \frac{1}{n^2}.$$

We can now express our inequality in these terms:

$$S_M - S_N \leq \frac{1}{N} - \frac{1}{M} \leq S_{M-1} - S_{N-1}$$

Next, let $M \to \infty$, so $S_M \to S$, and $1/M \to 0$. We have

$$S - S_N \leq \frac{1}{N} \leq S - S_{N-1}$$

Replacing N by $N + 1$, gives another inequality, which also holds, namely

$$S - S_{N+1} \leq \frac{1}{N+1} \leq S - S_N,$$

and combining these two, we have

$$S - S_{N+1} \leq \frac{1}{N+1} \leq S - S_N \leq \frac{1}{N} \leq S - S_{N-1}.$$

In particular, we have both upper and lower bounds for $S - S_N$, as

$$\frac{1}{N+1} \leq S - S_N \leq \frac{1}{N}$$

To make the point that this is a useful statement, we now specialise to the case when $N = 10$. Then

$$\frac{1}{11} \leq S - S_{10} \leq \frac{1}{10}$$

or

$$0 \leq S - \left(S_{10} + \frac{1}{11}\right) \leq \frac{1}{10} - \frac{1}{11} = \frac{1}{110}.$$

Our conclusion is that although S_{10} is not a very good approximation, we can describe the error well enough to get a much better approximation.

POWER SERIES

Power Series and the Radius of Convergence

The idea of writing $f(x) = P_n(x) + R_n(x)$, to express a function in terms of its Taylor polynomial, together with a remainder. For some functions, the remainder $R_n(x) \to 0$ as $n \to \infty$ for each fixed x. We now recognise this as showing that certain series converge.

We have more effective ways of showing that such a series converges we can use that ratio test. But note that such a test will only show that a series converges, not that it converges to the function used to generate it in the first place.

Proposition: The following series converge for all values of x to the functions shown:

$$e^x = 1 + x + \frac{x^2}{2!} + \frac{x^2}{2!} + \ldots \frac{x^n}{n!} + \ldots$$

$$\sin x = x - \frac{x^3}{3!} + \frac{x^5}{5!} + \ldots + (-1)^{n+1} \frac{x^{2n+1}}{(2n+1)!} + \ldots$$

$$\cos x = 1 - \frac{x^2}{2!} + \frac{x^4}{4!} + \ldots + (-1)^n \frac{x^{2n}}{2n!} + \ldots$$

$$\sinh x = x + \frac{x^3}{3!} + \frac{x^5}{5!} + \ldots + \frac{x^{2n+1}}{(2n+1)!} + \ldots$$

$$\cosh = 1 + \frac{x^2}{2!} + \frac{x^4}{4!} + \ldots + \frac{x^{2n}}{2n!} + \ldots$$

These are all examples of the subject of this section; they are real power series, which we can use to define functions. The corresponding functions are the best behaved of all the classes of functions; indeed are as well behaved as could possibly be expected. We shall see in this section that this class of functions are really just "grown up polynomials", and that almost any manipulation valid for polynomials remains valid for this larger class of function.

Definition: A real power series is a series of the form $\sum a_n x^n$, where the a_n are real numbers, and x is a real variable. We are thus dealing with a whole collection of series, one for each different value of x. Our hope is that there is some coherence; that the behaviour of series for different values of x are related in some sensible way.

Example: The geometric series $\sum_{n=0}^{\infty} x^n$ is another example of a power series we have already met. We saw this series is convergent for all x with $|x| < 1$.

It turns out that a power series is usually best investigated using the ratio test, Theorem. And the behaviour of power series is in fact very coherent.

Radius of Convergence

Theorem: Suppose $\sum a_n x^n$ is a power series. Then one of the following happens:

1. $\sum a_n x^n$ converges only when $x = 0$; or
2. $\sum a_n x^n$ converges absolutely for all x; or
3. There is some number $R > 0$ such that $\sum a_n x^n$ converges absolutely for all x with $|x| < R$, and diverges for all x with $|x| > R$.

No statement is made in the third case about what happens when $x = R$.

Definition: The number R described above is called the radius of convergence of the power series. By allowing $R = 0$ and $R = \infty$, we can consider *every* power series to have a radius of convergence.

Thus every power series has a radius of convergence. We

sometimes call the interval $(-R, R)$, where the power series is guaranteed to converge, the interval of convergence. It is characterised by the fact that the series converges (absolutely) inside this interval and diverges outside the interval.

1. The word "radius" is used, because in fact the same result is true for complex series, and then we have a genuine circle of convergence, with convergence for all (complex) z with $|z| < R$, and guaranteed divergence whenever $|z| > R$.

2. Note the power of the result; we are guaranteed that when $|x| > R$, the series diverges; it can't even converge "accidentally" for a few x's with $|x| > R$. Only on the circle of convergence is there ambiguity.

This regularity of behaviour makes it easy to investigate the radius of convergence of a power series using the ratio test.

Example: Find the radius of convergence of the series $\Sigma \dfrac{nx^n}{2^{n+1}}$.

Solution: Recall that the ratio test only applies to series of positive terms, so we look at the ratio of the moduli.

$$\frac{\left|(n+1)^{st} \text{ term}\right|}{\left|n^{th} \text{ term}\right|} = \frac{(n+1)|x|^{n+1}}{2^{n+2}} \frac{2^{n+1}}{n|x|^n}$$

$$= \frac{(n+1)|x|}{n.2} \to \frac{|x|}{2} \text{ as } n \to \infty.$$

Thus the given series diverges if $|x| > 2$ and converges absolutely if $|x| < 2$. Hence it has radius of convergence 2.

Example: Find the radius of convergence of the series $\Sigma \dfrac{(-1)^n n! x^n}{n^n}$.

Solution: This one is a little more subtle than it looks, although we have met the limit before. Again we look at the ratio of the moduli of adjacent terms.

$$\frac{\left|(n+1)^{st}\text{ term}\right|}{\left|n^{th}\text{ term}\right|} = \frac{(n+1)!|x|^{n+1}}{(n+1)^{n+1}}\frac{n^n}{n!|x|^n} = (n+1)\cdot|x|\cdot\frac{n^n}{(n+1)^{(n+1)}}$$

$$= \frac{n^n}{(n+1)^n}|x| = \left(\frac{n}{n+1}\right)^n|x|$$

$$= \frac{|x|}{\left(\frac{n+1}{n}\right)^n} \frac{|x|}{e}$$

as $n \to \infty$

Here we have of course used the result about e given in Section to note that

$$\left(1+\frac{1}{n}\right)^n \to e \text{ as } n \to \infty .$$

Thus the given series diverges if $|x| > e$ and converges absolutely (and so of course converges) if $|x| < e$. Hence it has radius of convergence e.

Exercise: Find the radius of convergence of the series

$$\sum_{n=0}^{\infty}\frac{x^n}{n^2+1}$$

We noted that the theorem gives no information about what happens when $x = R$, i.e. on the circle of convergence. There is a good reason for this — it is quite hard to predict what happens. Consider the following power series, all of which have radius of convergence 2.

$$\sum_{n=0}^{\infty}\frac{x^n}{2^n} \quad \sum_{n=1}^{\infty}\frac{x^n}{n2^n} \quad \sum_{n=1}^{\infty}\frac{x^n}{n^2 2^n}.$$

The first is divergent when $x = 2$ and when $x = -2$, the second converges when $x = -2$, and diverges when $x = 2$, while the third converges both when $x = 2$ and when $x = -2$. These results are all easy to check by direct substitution, and using Theorem.

Representing Functions by Power Series

Once we know that a power series has a radius of convergence, we can use it to define a function.

Suppose the power series $\sum a_n x^n$ has radius of convergence $R > 0$, and let $I = (-R, R)$. We now define a function f on this open interval I as follows:

$$f(x) = \sum_{n=1}^{\infty} a_n x^n \quad \text{for } x \in I.$$

In fact all of what we say below remains true when $R = \infty$, provided we interpret the open interval I as $\mathbb{R}$.

Theorem: Let $\sum a_n x^n$ be a power series with radius of convergence $R > 0$.

Let I be the open interval $(-R, R)$, and define $f(x) = \sum a_n x^n$ for $x \in I$. Then $\sum n a_n x^{n-1}$ has radius of convergence R, f is differentiable on I, and

$$f'(x) = \sum_{n=1}^{\infty} n a_n x^{n-1} \quad \text{for } x \in I.$$

We summarise this result by saying that we can differentiate a power series term – by - term everywhere inside the circle of convergence. If $R = \infty$, then this can be done for all x.

Proof: Quite a lot harder than it looks; we need to be able to re-arrange power series, and then use the Mean Value Theorem to estimate differences, and show that even when we add an infinite number of errors, they don't add up to too much.

Corollary: Let f and I be defined. Then f has an indefinite integral defined on I, given by

$$G(x) = \sum_{n=0}^{\infty} \frac{a_n}{(n+1)} x^{n+1} \quad \text{for } x \in I.$$

Other Power Series

We now derive some further power series to add to the collection described. Starting with the geometric series

$$\frac{1}{1-x} = 1 + x + x^2 + x^3 + ... + x^n + ... \quad \text{for } |x| < 1,$$

we replace x by $-x$ to get

$$\frac{1}{1+x} = 1 - x + x^2 - x^3 + ... + (-1)^n x^n + ... \quad \text{for } |x| < 1.$$

Integrating both sides then gives

$$\log(1+x) = K + x - \frac{x^2}{2} + \frac{x^3}{3} + ... + (-1)^n \frac{x^{n+1}}{n+1} + ... \quad \text{for } |x| < 1,$$

where K is a constant of integration. Evaluating both sides when $x = 0$ shows that $K = 0$, and so we get the series

$$\log(1+x) = \sum_{n=0}^{\infty} \frac{(-1)^n x^{n+1}}{n+1} \quad \text{valid for } |x| < 1$$

We return to equation, and replace x by $-x^2$ to get

$$\frac{1}{1+x^2} = 1 - x^2 + x^4 - x^6 + ... + (-1)^n x^{2n} + ... \quad \text{for } |x| < 1.$$

Again integrating both sides, we have arctan $(x) = K + x -$

$$+ ... + (-1)^n \frac{x^{2n+1}}{2n+1} + ... \quad \text{for } |x| < 1, \text{ where } K \text{ is a}$$

constant of integration. Again putting $x = 0$ shows that $K = 0$, and so

$$\arctan(x) = \sum_{n=0}^{\infty} \frac{(-1)^n x^{2n+1}}{2n+1} \quad \text{valid for } |x| < 1.$$

Example: Find the radius of convergence R of the power series

$$\frac{x^2}{2} - \frac{x^3}{3.2} + \frac{x^4}{4.3} - \frac{x^5}{5.4} .. + \frac{(-1)^n x^n}{n(n-1)} + ...$$

By differentiation or otherwise, find the sum of the series for $|x| < R$. [You may assume, without proof, that

$$\int \log(1+x)\,dx = K + x\log(1+x) - x + \log(1+x),$$

for some constant of integration K.]

Solution: Apply the ratio test to the given power series. Then

$$\left|\frac{a_n+1}{a_n}\right| = \left|\frac{x^{n+1}n(n-1)}{x^n n(n+1)}\right| \to |x| \quad \text{as } n \to \infty .$$

Thus the new series has radius of convergence 1. Denote its sum by $f(x)$, defined for $|x| < 1$.

Inside the circle of convergence, it is permissible to differentiate term - by - term, and thus $f'(x) = \log(1+x)$ for $|x| < 1$, since they have the same power series. Hence

$$f(x) = \int \log(1+x)\,dx = K + x\log(1+x) - \int \frac{x+1-1}{1+x}\,dx$$
$$= K + x\log(1+x) - x + \log(1+x).$$

Putting $x = 0$ shows that $K = 0$ and so $f(x) = (1+x)\log(1+x) - x$.

We have now been able to derive a power series representation for a function without working directly from the Taylor series, and doing the differentiations — which can often prove very awkward. Nevertheless, we have still found the Taylor coefficients.

Proposition: Let $\sum a_n x^n$ be a power series, with radius of convergence $R > 0$, and define

$$f(x) = \sum_{n=0}^{\infty} a_n x^n \quad \text{for } |x| < R.$$

Then $a_n = \dfrac{f^{(n)}(0)}{n!}$, so the given series *is* the Taylor (or Maclaurin) series for f

Proof: We can differentiate n times and we still get a series with the same radius of convergence. Also, calculating exactly as in the start of Section, we see that the derivatives satisfy $f^{(n)}(0) = n!a_n$, giving the uniqueness result.

Example: Let $f(x) = \frac{1}{1-x^3}$. Calculate $f^{(n)}(0)$.

Solution: We use the binomial theorem to get a power series expansion about 0,

$$\frac{1}{1-x^3} = 1 + x^3 + x^6 + x^9 + ... + x^{3n} + ... \quad \text{valid for } |x| < 1.$$

We now read off the various derivatives. Clearly $f^{(n)}(0) = 0$ unless n is a multiple of 3, while $f^{(3k)}(0) = (3k)!$

Power Series or Function

We have now seen that when a power series is used to define a function, then that function is *very* well behaved, and we can manipulate it by manipulating, in the obvious way, the corresponding power series. However there are snags. A function has *one* definition which works everywhere it makes sense (at least for simple functions), whereas the power series corresponding to a function depends also on the point about which the expansion is happening. An example will probably make this clearer than further discussion.

Example: Give the power series expansions for the function $f(x) = \frac{1}{1-x}$.

Solution: We can already do this about 0 by the Binomial Theorem; we have:

$$\frac{1}{1-x} = 1 + x + x^2 + x^3 + ... + x^n + ... \quad \text{for } |x| < 1.$$

To expand about a different point, e.g. about 3, write $y = x - 3$. Then

$$\frac{1}{1-x} = \frac{1}{1-(y+3)} = = \frac{1}{-2-y} = -\frac{1}{2} \cdot \frac{1}{1-y/2},$$

and again using the Binomial Theorem on the last representation, we have

$$\frac{1}{1-x} = -1\ \frac{1}{2}(1+ y/2 + y^2/4 + y^3/9 + ... + y^n/2^n + ...) \text{ for } |y/2| < 1.$$

Writing this in terms of x gives

$$\frac{1}{1-x} = -\frac{1}{2}\left(1+\frac{x-3}{2}+\frac{(x-3)^2}{4}+...+\frac{(x-3)^n}{2^n}+...\right)$$

for $|x-3| < 2$.

It should be no surprise that this is the Taylor series for the same function about the point 3. And it is in fact not an accident that the radius of convergence of the new series is 2. More investigation (quite a lot more - mainly for complex functions) shows the radius of convergence is always that of the *largest* circle that can be fitted into the domain of definition of the function. And that is why it is of interest to sometimes consider power series as complex power series. The power series expansion for $(1 + x^2)^{-1}$ has radius of convergence 1. This seems implausible viewed with real spectacles, but totally explicable when it is realised that the two points i and $-i$ are stopping the expansion from being valid in a larger circle.

THE DEFINITION AND LIMIT METHOD

The derivative of a function expresses its rate of change with respect to an independent variable. The derivative is also the slope of the tangent line to the curve. As you might recall. Many students start out differentiating with the Limit Method. The Limit Method utilizes the definition of a derivative of a function. The Limit Method, as seen below, is imperative because some problems may not work or may take longer with the differentiation rules.

$$\lim_{c\to 0}\frac{f(x+c)-f(x)}{c}$$

(Where c denotes a small distance on the x-axis usually written as the lower case delta sign next to an x)

DERIVATIVE RULES

The Constant Rule

The derivative of a constant function is 0. That is, if c is a real number, then d/dx[c] = 0.

The Sum and Difference Rules

The sum(or difference) of two differentiable functions is differentiable and is the sum(or difference) of their derivatives.

$$\frac{d}{dx}[f(x)+g(x)] = f'(x)+g'(x)$$

$$\frac{d}{dx}[f(x)-g(x)] = f'(x)-g'(x)$$

The Constant Multiple Rule

If f is a differentiable function and c is a real number, then cf is also differentiable and d/dx[$cf(x)$] = $cf'(x)$

The Power Rule

If n is a rational number, then the function $f(x) = x^n$ is differentiable and d/dx[x^n] = nx^{n-1}

The Product Rule

The product of two differentiable functions, f and g, is itself differentiable. Moreover, the derivative of fg is the first function times the derivative of the second, plus the second function times the derivative of the first.

$$\frac{d}{dx}[f(x)g(x)] = f(x)g'(x) + g(x)f'(x)$$

The Quotient Rule

The quotient f/g, of two differentiable functions, f and g, is itself differentiable at all values of x for which $g(x)$ does not = 0. Moreover, the derivative of f/g is given by the denominator times the derivative of the numerator minus the numerator times the derivative of the denominator divided by the square of the denominator.

$$\frac{d}{dx}\left[f(x)/g(x)\right] = (g(x)f'(x) - f(x)g'(x))/[g(x)]^2$$

g(x) does not = 0

The Chain Rule

If $y = f(u)$ is a differentiable function of u and $u = g(x)$ is a differentiable function of x, then $y = f(g(x))$ is a differentiable function of x and $d/dx[f(g(x))] = f'(g(x))g'(x)$

The General Power Rule

If $y = [u(x)]^n$, where u is a differentiable function of x and n is a rational number, then $d/dx = [u^n] = nu^{n-1}u'$.

IMPLICIT DIFFERENTIATION

An implicit function of x and y is a function in which one of the variables is not directly expressed in terms of the other. In other words, you use implicit differentiation when both x and y are on the same side. To implicitly differentiate an implicit function follow the steps below.

1. Apply all differentiation rules. Except to y.
   $y^3 - x^2 = -5$
   $y^3 - 2x = 0$
2. Then differentiate the y's in regard to x. In turn you add a y prime to the differentiated y's.
   $3y^2y' - 2x = 0$
3. Now factor out any y primes. Hint. In this example there are none.
4. Lastly, place all the y primes on one side.
   $y' = (2x) / (3y^2)$

TRIGONOMETRIC DIFFERENTIATION

Trigonometric derivatives frequently seen in the Calc AP AB

Test. These derivatives will never go away! As a result, every well prepared student knows the derivatives below by heart.

Trigonometric Derivatives:

$d/dx(\sin x) = \cos x$
$d/dx(\cos x) = -\sin x$
$d/dx(\tan x) = (\sec x)^2$
$d/dx(\sec x) = (\sec x)^2(\tan x)$
$d/dx(\csc x) = -(\csc x)^2(\cot x)$
$d/dx(\cot x) = -(\csc x)^2$

Notice that all co functions are negative.

Inverse Trigonometric Differentiation

Unlike the six basic trigonometric functions, the six inverse trigonometric functions must have restricted domains. The reason being that the six basic trigonometric functions are periodic and as a result have no on-to-one inverses.

Inverse Trigonometric Functions

Function	*Domain*	*Range*
y = arcsin x iff sin $y = x$	$-1 < x < 1$	$-\pi/2 < y < \pi/2$
y = arccos x iff cos $y = x$	$-1 < x < 1$	$0 < y < \pi$
y = arctan x iff tan $y = x$	–infinity $< x <$ infinity	$-\pi/2 < y < \pi/2$
y = arccot x iff cot $y = x$	–infinity $< x <$ infinity	$0 < y < \pi$
y = arcsec x iff sec $y = x$	$\lvert x \rvert > 1$	$0 < y < \pi, y \neq 0$
y = arccsc x iff csc $y = x$	$\lvert x \rvert > 1$	$-\pi/2 < y < \pi/2, y \neq 0$

Higher Ordered Derivatives

A higher ordered derivative is nothing more than a second derivative or greater. In other words, it is a function that was differentiated more than once. Why would we differentiate again? Well, higher ordered derivatives are usually used in finding the acceleration from teh position function, as seen below.

Position function = $s(t)$
Velocity function = $s'(t)$
Acceleration function = $s''(t)$

Exponential and Logarithmic Differentiation

Exponential Functions to Base a

If a is a positive real number (a does not = 1) and x is any real number, then the exponential function to the base a is denoted by a^x and is defined as $^x = e^{(\ln a)x}$.

If $a = 1$, then $y = 1^x = 1$ is a constant function.

Logarithmic Functisons to Base a

If a is a positive real number (a does not = 1) and x is any real number, then the logarithmic function to the base a is denoted by $\log_a x$ and is defined as $\log_a x = (1/\ln a) \ln x$

Derivatives for Bases other than e

Let a be a positive real number (a does not = 1) and let u be a differentiable function of x.

$$\frac{d}{dx}[a^x] = (lna)a^x$$

$$\frac{d}{dx}[a^u] = (ln\ a)a^u \text{ du/dx}$$

$$\frac{d}{dx}[\text{lob}_a x] = 1/((ln\ a)x)$$

$$\frac{d}{dx}[\log_a u] = 1/((ln\ a)u) \text{ du/dx}$$

The Power Rule for Real Exponents

Let n be any real number and let be a differentiable function of x.

$$\frac{d}{dx}[x^n] = \quad nx^{n-1}$$

$$\frac{d}{dx}[u^n] = \quad nu^{n-1} \text{ du/dx}$$

Derivative of the Natural Logarithmic Function

Let u be a differentiable function of x.

1. $\frac{d}{dx}[\ln x] = 1/x,\ x > 0$

But don't forget the Chain rule!

2. $\frac{d}{dx}[\ln u] = u'/u,\ u > 0$

The Mean Value Theorem

The mean value theorem refers to the mean or average rate of the change of f in the interval $[a,b]$. Listed below are two variations of the mean value theorem. The first is the basic theorem and the second is an extended version.

The Mean Value Theorsem

If f is continuous on the closed interval $[a,b]$ and differentiable on the open interval (a,b), then there exists a number c in (a,b) such that $f'(c) = (f(b) - f(a)) / (b - a)$

The Extended Mean Value Theorem

If f and g are differentiable on the open interval (a, b) and continuous on $[a\ b]$ such that $g'(x)$ does not $= 0$ for any x in (a, b), then there exists a point c in (a, b) such that $f'(c)/g'(c) = (f(b) - f(a)) / (g(b) - g(a))$

Additional Theorems and Rules of Differentiable Functions

If $f(x)$ is differentiable at c, it is continuous at c.

The Extreme Value Theorem

If $f(x)$ is continuous on the closed interval $[a, b]$, then f has both a minimum and maximum on the interval.

Relative Extrema Occur only at Critical Numbers

If f has a relative minimum or relative maximum at $x = c$, then c is a critical number of f.

Rolle's Theorem

Let f be continuous on the closed interval $[a, b]$ and differentiable on the open interval (a, b). If $f(a) = f(b)$ then there is at least one number c in (a, b) such that $f'(c) = 0$.

Absolute Value Rule

If u is a differentiable function of x such that u does not = 0, then $d/dx[\ln | u |] = u'/u$

L'Hopital's Rule

Let f and g be functions that are differentiable on an open interval (a, b) containning c, except possibly at c itself. Assume that $g'(x)$ does not = 0 for all x in (a, b), except possibly at c itself. If the limit of $f(x)/g(x)$ as x approaches c produces the indeterminate for 0/0, then

$$\lim_{x \to c} \frac{f(x)}{g(x)} = \lim_{x \to c} \frac{f'(x)}{g'(x)}$$

provided the limit on the right exists or is infinite. This result also appiies if the limit of $f(x)/g(x)$ as x approaches c produces any one of the indeterminate forms infinity/infinity, (–infinity)/infinity, infinity/(–infinity), or (–infinity)/(–infinity). Note that you do not apply the quotient rule! The rule above utilizes $f'(x)/g'(x)$ and not the derivative of $f(x)/g(x)$.

7

Principles of Differentiation

The natural method of constructing the integral and differential calculus is first to learn to differentiate the widest possible classes of functions and then, by virtue of the fundamental theorem, make the results thus obtained available for the solution of integration problems. In the following sections, it will be our task to carry out this programme.

In this development of the subject, certain rules for differentiation, with the first of which we are already acquainted, will have an important role. We assume that in the interval under consideration the functions $f(x)$ and $g(x)$ are differentiable; our rules then are:

Rule 1: Derivative of a Sum

If $\phi(x) = f(x) + g(x)$, then $\phi(x)$ is differentiable, and

$$\phi'(x) = f'(x) + g'(x)$$

that is, the processes of differentiation and addition are interchangeable. The same holds for the sum of any finite number n of terms

$$\phi(x) = \sum_{\nu-1}^{n} f_\nu(x),$$

for which we obtain

$$\phi'(x) = \sum_{\nu-1}^{n} f'_\nu(x).$$

We may pass over the proof, which, is fairly obvious.

Rule 2: Derivative of a Product

If $\phi(x) = f(x)g(x)$, then $\phi(x)$ is differentiable and

$$\phi'(x) = f(x)+g'(x)+g(x)+f'(x)$$

The proof follows from the equation

$$\frac{\phi(x+h)-\phi(x)}{h} = \frac{f(x+h)g(x+h)-f(x)g(x)}{h}$$

$$= \frac{f(x+h)g(x+h)-f(x+h)g(x)}{h}$$

$$+\frac{f(x+h)g(x)-f(x)g(x)}{h}$$

$$= f(x+h)\frac{g(x+h)-g(x)}{h}$$

$$+g(x)\frac{f(x+h)-f(x)}{h}$$

The passage to the limit $h \to 0$ can here be carried out directly and yields the result shown. This formula takes a still more elegant form, if we divide throughout by $\phi(x)=f(x)g(x)$. We then obtain

$$\frac{\phi'(x)}{\phi(x)} = \frac{f'(x)}{f(x)}+\frac{g'(x)}{g(x)}$$

where we, of course, assume that $\phi(x)$ is nowhere equal to zero. By repeated application of this product formula, we obtain, by induction, for the derivative of a product of n factors an expression consisting of n terms, each of which consists of the derivative of one factor multiplied by all the other factors of the original product. In symbols, this is

$$\phi'(x) = \frac{d}{dx}\{f_1(x)f_2(x).....f_n(x)\}$$

$$= f_1'(x)f_2(x)....f_n(x)$$

$$+f_1(x)f_2'(x)f_3(x)....f_n(x)$$

$$+....+f_1(x)f_2(x).....f_n'(x)$$

$$= \sum_{\nu-1}^{n} f_{\nu}'(x) \frac{\phi(x)}{f_{\nu}(x)},$$

or on division by

$$\phi(x) = f_1(x) f_2(x) \ldots f_n(x),$$

assuming again that $\phi(x)$ does not vanish anywhere,

$$\frac{\phi'(x)}{\phi(x)} = \frac{f_1'(x)}{\phi(x)} + \frac{f_2'(x)}{\phi(x)} + \ldots + \frac{f_n'(x)}{f_n(x)}$$

$$= \sum_{\nu=1}^{n} \frac{f_{\nu}'(x)}{f_{\nu}(x)}$$

Rule 3: Derivative of a Quotient

We have for a quotient

$$\phi(x) = \frac{f(x)}{g(x)}$$

the rule: The function $\phi(x)$ is differentiable at every point at which $g(x)$ does not vanish and

$$\phi'(x) = \frac{g(x) f'(x) - g'(x) f(x)}{\{g(x)\}^2}.$$

If $\phi(x) < 0$, this can be rewritten

$$\frac{\phi'(x)}{\phi(x)} = \frac{f'(x)}{f(x)} - \frac{g'(x)}{g(x)}.$$

If we accept the differentiability of $\phi(x)$ as a hypothesis, we can apply the product rule to,

$$f(x) = f(x) g(x)$$

and conclude that

$$f'(x) = \phi(x) g'(x) + g(x) \phi'(x)$$

By substituting $f(x)/g(x)$ for $\phi(x)$ on the right hand side and solving for $\phi'(x)$, we obtain the above rule. In order to prove the differentiability of $\phi(x)$ as well as the rule, we use the method: We write

$$\frac{\phi(x+h)-\phi(x)}{h} = \frac{\frac{f(x+h)}{g(x+h)}-\frac{f(x)}{g(x)}}{h}$$

$$= \frac{g(x)\frac{f(x+h)-f(x)}{h}-\frac{g(x+h)-g(x)}{h}f(x)}{g(x)g(x+h)}$$

If we now let h tend to 0, we arrive at the result stated, because, by assumption, the two terms obtained by performing the division on the right hand side have definite limits, which are

$$\frac{g(x)f'(x)}{\{g(x)\}^2} \text{ and } \frac{g'(x)f(x)}{\{g(x)\}^2},$$

respectively. This proves immediately the existence of the limit on the left hand side as well as the differentiation formula.

Rule 4: Multiplication by a Constant

If c is a constant and $\phi(x) = cf(x)$, then $\phi(x)$ is differentiable and

$$\phi'(x) = cf'(x)$$

This follows immediately from the relation

$$\frac{\phi(x+h)-\phi(x)}{h} = c\frac{f(x+h)-f(x)}{h}$$

if we take the limit $h \to 0$.

Differentiation of the Rational Functions

To begin with, we shall again deduce the differentiation

formula,

$$\frac{d}{dx}x^n = nx^{n-1}$$

for every positive integer n, basing the proof on the rule for differentiation of a product. x and hence obtain

$$\frac{d}{dx}x^n = 1.x^{n-1} + 1.x^{n-1} + \ldots + 1.x^{n-1}$$
$$= nx^{n-1}.$$

If we use this formula and the first rule of differentiation, the second derivative of the function x^n becomes:

$$\frac{d^2}{dx^2}x^n = n(n-1)x^{n-2}$$

Continuing this process, we obtain

$$\frac{d^3}{dx^3}x^n = n(n-1)(n-2)x^{n-3}$$

$$\frac{d^3}{dx^3}x^n = 1.2 \ldots n = n!$$

It is clear from the last of these formulae that the $(n + 1)$-th derivative of x^n vanishes everywhere. By virtue of our first two rules, a knowledge of the differentiation of powers enables us at once to differentiate any polynomial

$$y = a_0 + a_1 x + a_2 x + \ldots + a_n x^n.$$

We have simply

$$y' = a_1 + 2a_2 x + 3a_3 x^2 + \ldots + na_n x^{n-1}$$

and

$$y'' = 2a_2 + 3.2a_3 x + 4.3a_4 x^2 + \ldots + n(n-1)a_n x^{n-2}$$

and so on.

The differentiation of any rational function now follows with the help of the quotient rule. In particular, we shall again derive the differentiation formula for the function x^n, where $n = -m$ is a negative integer. The application of the quotient rule together with the fact that the derivative of a constant is equal to zero yields

$$\frac{d}{dx}\left(\frac{1}{x^m}\right) = -\frac{mx^{m-1}}{x^{2m}}$$

$$= -\frac{m}{x^{m+1}}$$

or, if we take $m = -n$,

$$\frac{d}{dx}x^n = nx^{n-1}$$

which agrees formally with the result for positive values of n and with the results given earlier.

THE DERIVATIVES OF A FUNCTION OF SEVERAL VARIABLES

Definition. Geometrical Representation

If we assign in a function of several variables definite numerical values to all but one of the variables and allow only that one variable, say x, to vary, the function becomes a function of one variable. For

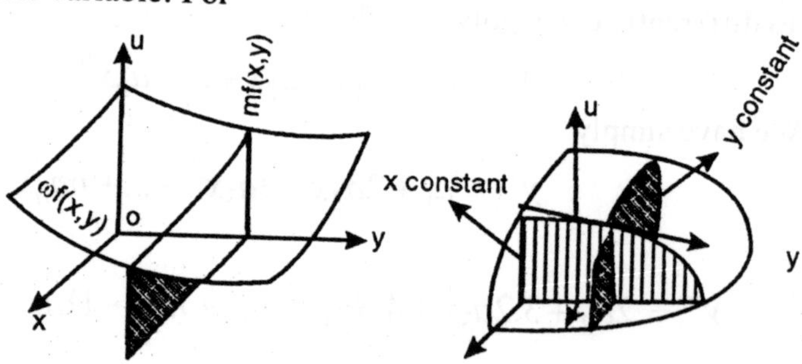

Section of u=f(x,y)

example, consider a function $u = f(x,y)$ of the two variables x and y and give y the definite fixed value $y = y_0 = c$.

The function $u = f(x, y_0)$ of the single variable x, which is thus formed, may be simply represented geometrically by letting the surface $u = f(x, y)$ be cut by the plane $y = y_0$. The curve of intersection thus formed in the plane is represented by the equation $u = f(x, y_0)$.

If we differentiate this function in the usual way at the point $x = x_0$ (assuming that the derivative exists), we obtain the partial derivative of $f(x,y)$ with respect to x at the point (x_0, y_0).

According to the ordinary definition of the derivative, this is the limit,

$$\lim_{h \to 0} \frac{f(x_0 + h, y_0) - f(x_0, y_0)}{h}$$

If (x_0, y_0) is a point on the boundary of the region of definition, we make the restriction that in the passage to the limit the point,

$$(x + h, y_0)$$

must always remain in the region. Geometrically speaking, this partial derivative denotes the tangent of the angle between a parallel to the x-axis and the tangent line to the curve $u = (x,y_0)$.

It is therefore the slope of the surface u = $f(x,y)$ in the direction of the x-axis. In order to represent these partial derivatives, several different notations are in use of which we mention:

$$\lim_{h \to 0} \frac{f(x_0 + h, y_0) - f(x_0, y_0)}{h} = f_\infty(x_0, y_0) = u_\infty(x_0, y_0)$$

If we wish to emphasize -that the partial derivative is the limit of a difference quotient, we denote it by

$$\frac{\partial f}{\partial x} or \frac{\partial}{\partial x} f .$$

We use here a special curved letter ∂ instead of the ordinary d used in the differentiation of functions of one variable, in order to show that we are dealing with a function of several variables and differentiating with respect to one of them.

Sometimes. it is convenient to use Cauchy's symbol D and to write,

$$\frac{\partial f}{\partial x} = D_{\infty} f;$$

however, we shall rarely use this symbol.

In exactly the same way, we define the partial derivative of $f(x, y)$ with respect to y at the point (x_0, y_0) by the relation

$$\lim_{h \to 0} \frac{f(x_0, y_0 + k) - f(x_0, y_0)}{k} = f_v(x_0, y_0)$$
$$= \frac{\partial f}{\partial y} = D_v f(x_0, y_0)$$

It represents the slope of the curve of intersection of the surface $u = f(x, y)$ with the plane $x = x_0$, perpendicular to the x-axis.

Now, let us think of the point (x_0, y_0), hitherto considered to be fixed, as variable and accordingly omit the suffices. In other word, we think of the differentiation as being carried out at any point (x, y) of the region of definition of $f(x, y)$. Then the two derivatives are themselves functions of x and y:

$$u_x(x, y) = f_x(x, y) = \frac{\partial f(x, y)}{\partial x} \text{ and}$$
$$u_y(x, y) = f_y(x, y) = \frac{\partial f(x, y)}{\partial y}$$

Similarly, we make the definition for any number (n) of independent variables:

$$\frac{\partial f(x_1, x_2, ..., x_n)}{\partial x_1} = \lim_{h \to 0} \frac{f(x_1 + h, x_2 ..., x_n)}{h}$$
$$\lim_{h \to 0} \frac{-f(x_1 + x_2 ..., x_n)}{h}$$
$$= fx_1(x_1, x_2, ..., x_n)$$
$$= Dx_1 f(x_1, x_2, ..., x_n),$$

it being assumed that the limit exists. Of course, we can also form higher partial derivatives of $f(x, y)$ by again differentiating the partial, first order derivatives $f_x(x, y)$ and $f_y(x, y)$ with respect to one of the variables, and repeating this process. We indicate the order of differentiation by the order of the suffixes or by the order of the symbols ∂x and ∂y in the denominator from the right to the left hand side, and use for the second partial derivatives the: On the other hand, in Continental usage, $\frac{\partial}{\partial x}\left(\frac{\partial f}{\partial y}\right)$ becomes $\frac{\partial^2 f}{\partial y \partial x}$

We likewise denote the third partial derivatives by

$$\frac{\partial}{\partial x}\left(\frac{\partial^2 f}{\partial x^2}\right) = \frac{\partial^2 f}{\partial x^3} = f_{xxx},$$

$$\frac{\partial}{\partial y}\left(\frac{\partial^2 f}{\partial x^2}\right) = \frac{\partial^3 f}{\partial y dx^2} = f_{yxx},$$

$$\frac{\partial}{\partial x}\left(\frac{\partial^2 f}{\partial x \partial y}\right) = \frac{\partial^2 f}{\partial x^2 \partial y} = f_{xxy}, \text{ etc.}$$

and, in general, the n-th derivatives

$$\frac{\partial}{\partial x}\left(\frac{\partial^{n-1} f}{\partial x^{n-1}}\right) = \frac{\partial^n f}{\partial x^n} = fx^n$$

$$\frac{\partial}{\partial y}\left(\frac{\partial^{n-1} f}{\partial x^{n-1}}\right) = \frac{\partial^n f}{\partial y \partial x^{n-1}} = fyx^{n-1}, \&c.$$

Finally, we shall study a few examples of the actual calculation of partial derivatives.

According to the definition, all the independent variables are to be kept constant except the one with respect to which we are differentiating. We therefore have merely to regard the other variables as constants and carry out out the differentiation according to the rules by which we differentiate functions of a single independent variable.

For example, we have:

1. Function $f(x, y) = xy$

   First Derivatives, $f_x = y, f_v = x$

   Second Derivatives, $f_{xx} = 0, f_{xy} = f_{yx} = 1, f_{yy} = 0$

2. Function $f(x,y) = \sqrt{(x^2 + y^2)}$

First Derivatives,

$$f_x = \frac{x}{\sqrt{(x^2+y^2)}}, f_v = \frac{y}{\sqrt{x^2+y^2}} \cdot r$$

(Thus for the radius vector $r = \sqrt{(x^2+y^2)}$ from the origin to the point (x, y), the partial derivatives with respect to x and y are given by the cosine, $\cos\varphi = x/r$, and the sine, $\sin\varphi = y/r$ of the angle φ, which the radius vector makes with the positive direction of the x-axis.)

Second Derivatives

$$f_{xx} = \frac{\sqrt{(x^2+y^2) - \frac{x^3}{\sqrt{(x^2+y^2)}}}}{x^2+y^2} = \frac{y^2}{\sqrt{(x^2+y^2)^2}} = \frac{\sin^2\varphi}{r}$$

$$f_{xy} = f_{yx} = -\frac{xy}{\sqrt{(x^2+y^2)^2}} = -\frac{\sin\varphi\cos\varphi}{r}$$

$$f_{yy} = \sqrt{(x^2+y^2) - \frac{y^2}{\sqrt{(x^2+y^2)}}} = \frac{x^2}{\sqrt{(x^2+y^2)^2}} = \frac{\cos^2\varphi}{r}$$

1. Reciprocal of the radius vector in three dimensions:

$$f(x, y, z) = \frac{1}{\sqrt{(x^2 + y^2 + z^2)}} = \frac{1}{r}$$

First Derivatives,

$$f_x = -\frac{x}{\sqrt{(x^2 + y^2 + z^2)^2}} = -\frac{x}{r^2}$$

$$f_y = -\frac{x}{\sqrt{(x^2 + y^2 + z^2)^2}} = -\frac{y}{r^2}$$

$$f_z = -\frac{x}{\sqrt{(x^2 + y^2 + z^2)^2}} = -\frac{z}{r^2}$$

Second Derivatives,

$$f_{xx} = -\frac{1}{r^2} + \frac{3x^2}{r^2}, f_{yy} = -\frac{1}{r^2} + \frac{3y^2}{r^2}, f_{zz} = -\frac{1}{r^2} + \frac{3z^2}{r^2}$$

$$f_{xy} = f_{yx} = \frac{3xy}{r^2}, f_{yz} = f_{zy} = \frac{3yz}{r^2}, f_{zx} = f_{zx} = \frac{3zx}{r^2}$$

We see from this that there holds for the function $f = \dfrac{1}{\sqrt{(x^2 + y^2 + z^2)}}$ the equation,

$$f_{xx} + f_{yy} + f_{zz} = -\frac{3}{r^2} + \frac{3(x^2 + y^2 + z^2)}{r^2} = 0,$$

for all values of x, y, z except 0,0,0; as we may say, the equation

$$f_{xx} + f_{yy} + f_{zz} = 0$$

is satisfied identically in x, y, z by the function $f(x, y, z) = 1/r$.

2. Functions, $f(x,y)=\frac{1}{\sqrt{y}}e^{-(x-a)^2/4y}$

First Derivatives,

$$f_x = \frac{1}{\sqrt{y}}\frac{-(x-a)}{2y}e^{-(x-a)^2/4y}$$

$$f_y = \left(\frac{-1}{2y^{3/2}}+\frac{(x-a)^2}{4y^{3/2}}\right)e^{-(x-a)^2/4y}$$

Second Derivatives,

$$f_{xx} = \left(\frac{-1}{2y^{3/2}}+\frac{(x-a)^2}{4y^{3/2}}\right)e^{-(x-a)^2/4y}$$

The equation

$$f_{xx}-f_y = 0$$

is therefore satisfied identically in x and y.

Just as in the case of one independent variable, the possession of derivatives is a special property of a function. Note that the expression differentiable implies more than that the partial derivatives with respect to x and y exist. All the same, this property is possessed by all functions of practical importance except perhaps at isolated exceptional points.

In contrast to functions of one variable, the possession of derivatives does not imply the continuity of the function. This is clearly shown by the example $u=\frac{2xy}{x^2+u^2}$ already considered. because the partial derivatives exist for it everywhere and yet the function is discontinuous at the origin. But, as is stated by the following theorem, the possession of bounded derivatives does imply continuity: If a function $f(x,y)$ has partial derivatives f_x and f_y everywhere in a region R, and these derivatives satisfy everywhere the in equalities

$$|f_x(x,y)| < M, |f_x(x,y)| < M,$$

where M is independent of x and y, then $f(x, y)$ is continuous everywhere in R. In particular, if f_x and f_y are continuous, they are necessarily bounded, so that $f(x,y)$ is also continuous. In other words, it made no difference whether we differentiated first with respect to x and then with respect to y or *vice versa*. This is no accidental occurrence.

Theorem: If the mixed partial derivatives f_{xy} and f_{yx} of a function $f(x,y)$ are continuous in a region R, then the equation

$$f_{yx} = f_{xy}$$

holds everywhere in the interior of this region, i.e., the order of differentiation with respect to x and y is immaterial.

Applying this theorem to f_x and f_y, then to f_{xx}, f_{xy}, f_{yy}, etc., we find that

$$f_{xxy} = f_{xyx} = f_{yxx},$$
$$f_{xyy} = f_{xyx} = f_{yyx},$$
$$f_{xxyy} = f_{xyxy} = f_{xyyx} = f_{yxxy} = f_{yxyx} = f_{yyxx}, \text{ \& c.,}$$

and, in general, we have the result:

In repeated differentiation of a function of two variables, the order of differentiation can be changed arbitrarily, provided only that the derivatives in question are continuous functions.

DIFFERENTIATION OF THE TRIGONOMETRIC FUNCTIONS

For the trigonometric functions sin x and cos x, we have already obtained the differentiation formulae

$$\frac{d}{dx}\sin x = \cos x$$

and

$$\frac{d}{dx}\cos x = -\sin x$$

The quotient rule now enables us to differentiate the functions,

$$y = \tan x$$
$$= \frac{\sin x}{\cos x} \text{ and}$$

$$y = \cot x$$
$$= \frac{\cos x}{\sin x}$$

According to the rule, the derivative of the first of these functions is

$$y' = \frac{\cos^2 x + \sin^2 x}{\cos^2 x}$$
$$= \frac{1}{\cos^2 x'}$$

and we obtain the result

$$\frac{d}{dx}\tan x = \frac{1}{\cos^2 x}$$
$$= \sec^2 x$$
$$= 1 + \tan^2 x$$

Similarly, we obtain

$$\frac{d}{dx}\cot x = -\frac{1}{\sin^2 x}$$
$$= -\cos ec^2 x$$
$$= -\left(1 + \cot^2 x\right).$$

THE DIFFERENTIATION OF INVERSE FUNCTIONS

Functions of Functions (Compound Functions)

It often happens that a function *u* of the independent variables

x, y is stated in the form,

$$u = f(\xi, \eta, \ldots),$$

where the arguments ξ, η,... of the function f are themselves functions of x and y:

$$\xi = \phi(x, y),\ \eta = \psi(x, y), \ldots$$

We then say that,

$$u = f(x, \eta, \ldots), f = (f(x, y), \ldots y(x, y), \ldots) = F(x, y)$$

is given as a compound function of x and y.

For example, the function

$$u = e^{x^3} y(x+y)^3$$

may be written as a compound function by means of the relations

$$u = e^{\xi}\eta^3 = f(\xi, \eta);\ \xi = x^2 y,\ \eta = x + y.$$

Similarly, the function

$$u = \log(x + 1).\ \text{arc cos}\ \sqrt{(4 - x^2 - y^2)}$$

can be expressed in the form

$$u = \eta \arccos \xi = f(\xi, \eta),$$

$$\xi = \sqrt{(4 - x^2 - y^2)}, \eta = \log(x+1).$$

In order to make this concept more precise, we assume to begin with that the functions $\xi = \phi(x, y)$, $\eta = \psi(x, y)$,... are defined in a certain region R of the independent variables x, y.

Then there corresponds to every point (x,y) of R a point $(\xi, \eta, ..)$. As the point (x,y) ranges over R, the point $(\xi, \eta, \ldots)$ will range over a certain set of values. We assume that the point $(\xi, \eta, \ldots)$ always lies within a region S in which $f(\xi, \eta, \ldots)$ is defined. The function

$$u = f(x, y), \psi(x, y), \ldots) = F(x, y)$$

is then defined in the region R.

Referring to our examples, we find in the first one of them that ξ and η are defined for every x, y and $f(\xi, \eta)$ is defined for every ξ, η, so that our region R can be taken to be the entire *xy*-

plane. However, in the second example, the region S is restricted by the inequality $|\xi| \le 1$, since for $|\xi| > 1$ the function across ξ is undefined.

Secondly, the region R is restricted by the inequalities $x+1>0$ and $x + y \le 4$, since for other values both ξ and η are not defined.

Thirdly, the region R must be further limited by the inequality $3 \le x + y$ in order that the point with co-ordinates ξ, η shall fall into S, i.e., the restriction $|\xi| \le 1$ implies that $x + y \le 3$. Hence R consists of the part of the ring $3 \le x+y \le 4$ lying to the right of the line $x = -1$.

The following theorem on compound functions is an immediate consequence of the definitions: If the function $u = f(\xi, \eta, \ldots)$ is continuous in S and the functions $\xi = \phi(x, y)$, $\eta = \psi(x, y)$), are continuous in R, then the compound function $u = F(x, y)$ is continuous in $R.\xi = \phi\ (x, y)$. The reader should be able to prove this statement

The Chain Rule

We now turn our attention to compound functions of the type $u = f(\xi, \eta, \ldots)$, where $\xi, \eta, \ldots$ depend on the single variable x:

$$\xi = \phi(x),\ \eta = \psi(x), \ldots.$$

For such functions, we have the important theorem known as the chain rule: If the function $u = f(\xi, \eta, \ldots)$ has continuous partial derivatives of the first order in S and the functions $\xi = \phi(x)$, $\eta = \psi(x), \ldots$ have continuous first derivatives in the interval R, $a \le x \le b$, then $u = f(\phi(x), \psi(x), \ldots) = F$ xi) has a continuous derivative in R and

$$F'(x) = f_\xi \phi'(x) + f_\eta \eta'(x) + \ldots.$$

The right-hand side of this equation is an abbreviation for

$$f_x\ \{f(x),\ y(x), \ldots\}\ f'(x) + \ldots$$

In order to simplify the notation, we shall assume that f is a function of the three arguments ξ, η, ζ.

We shall denote by x_0 an arbitrary fixed point of the interval $a \le x \le b$, by ξ_0, η_0, ζ_0 the corresponding values $\xi_0 = \phi(x_0)$, $\eta_0 = \psi(x_0)$, $\xi_0 = \chi(x_0)$ and by ξ, η, ζ the values $\phi(x)$, $\psi(x)$, $\chi(x)$, corresponding to a variable point $x = x_0 + h$. We first write down the identity,

$$F(x) - F(x_0) = f(\xi,\eta,\zeta) - f(\xi_0,\eta_0,\zeta_0)$$

$$= f(\xi,\eta,\zeta) - f(\xi_0,\eta_0,\zeta_0)$$

$$+\{f(\xi_0,\eta,\zeta) - f(\xi_0,\eta_0,\zeta)\}$$

$$= \{f(\xi_0,\eta_0,\zeta) - f(\xi_0,\eta_0,\zeta_0)\}$$

In each bracket on the right hand side, we observe that only one of the independent variables changes its value. Hence we can apply to each bracket the mean value theorem for functions of a single variable and obtain

$$F(x) - F(x_0) = (\xi - \xi_0) f_\xi(\overline{\xi},\eta,\xi) + (\eta - \eta_0) f_\eta(\xi_0,\overline{\eta},\zeta)$$

$$+(\xi - \xi_0) f_\zeta(\xi_0,\eta_0,\overline{\zeta}),$$

where $\overline{\xi}$ lies between x and x_0, $\overline{\eta}$ between h_0 and h, and $\overline{\zeta}$ between z_0 and z. Moreover, by the mean value theorem, we have

$$\xi - \xi_0 = \phi(x) - \phi(x_0) = (x - x_0)\phi'(x_1),$$

$$\eta - \eta_0 = \psi(x) - \psi(x_0) = (x - x_0)\psi'(x_2),$$

$$\zeta - \zeta_0 = \chi(x) - \chi(x_0) = (x - x_0)\chi'(x_3),$$

where all x_1, x_2, and x_3 lie between x_0 and x. Substituting these values in the last equation and dividing by $x - x_0$, we have

$$\frac{F(x) - F(x_0)}{x - x_0} = f_\xi(\overline{\xi},\eta,\zeta)\phi'(x_1) + f_\eta(\xi_0,\overline{\eta},\zeta)\psi'(x_2)$$

$$+ f_\zeta(\xi_0,\eta_0,\overline{\zeta})\chi'(x_3)$$

We now let x tend to x_0. Owing to the continuity of $\phi(x)$, $\psi(x)$, $\chi(x)$, the quantities ξ,η,ζ tend to ξ_0,η_0,ζ_0, respectively, and, *a fortiori*, so do $\overline{\xi},\overline{\eta},\overline{\zeta}$. Also x_1, x_2 and x_3 tend to x_0. Since all the functions on the right hand side are continuous, we have

$$\lim_{n \to x_0} \frac{F(x) - F(x_0)}{x - x_0} = F'(x_0) = f_\xi(\xi_0,\eta_0,\zeta_0)\phi'(x_0)$$

$$+f_\eta(\xi_0,\eta_0,\zeta_0)\psi'(x_0)$$
$$+f_\zeta(\xi_0,\eta_0,\zeta_0)\chi'(x_0),$$

thus establishing the formula for $F'(x)$.

The continuity of $F'(x)$ follows immediately from the formula, since, by assumption, ϕ', ψ', χ', are continuous f_ξ, f_η and f_ζ are are continuous functions of continuous functions. This theorem may be extended to compound functions of two or more variables, as follows:

If the funding $u = f(\xi, \eta, ...)$ has continuous partial derivatives of the first order in the region S and the functions $\xi = \phi(x,y)$, $\eta = \psi(x,y)$, x,y have continuous partial derivatives of the first order in R, then $u = F(x, y) = f\{\phi(x,y), \psi(x,y), ...$ has continuous partial derivatives of the first order in R and their derivatives are given by

$$F_x = f_\xi\phi_x + f_\eta\psi_x + ...,$$
$$F_y = f_\xi\phi_y + f_\eta\psi_y + ...$$

These formulae are often written in the abbreviated form

$$u_x = u_\xi\xi_x + u_\eta\psi_x + ...,$$
$$u_y = u_\xi\xi_y + u_\eta\eta_y + ...$$

In order to derive them, we temporarily introduce the notation $g(x) = \phi(x,y_0)$, $h(x) = \psi(x,y_0), ...,$ where y_0 is a fixed value of y. By the definition of the partial derivatives, it follows that $g'(\mathrm{x}) = \phi_x(x,y_0)$, $h'(\mathrm{x}) = \psi_x(x,y_0)$,. Similarly, if we writs $H(\mathrm{x}) = F(x, y_0)$, we have $H'(x) = F_x(x, y_0)$. We now apply the theorem just proved to the function $u = H(x) = f(\xi, \eta, ...) = f\{g(x), h(x), ...\}$ and obtain

$$H'(x_0) = f_\xi g'(x_0) + f_\eta h'(x_0) + ...$$

Returning to the original symbols, we have

$$F_x(x_0, y_0) = f_\xi\phi_x(x_0, y_0) + f_\eta\psi_x(x_0, y_0) + ...$$

The other formula is proved in a similar manner.

If we wish to calculate higher order derivatives, we need only differentiate again the right hand side of our formulae with respect to x and y, regarding $f_\xi, f_\eta, ...$ as compound functions. Thus, for u =

$f(\xi,\eta) = f\{\phi(x,y),\ \psi(x,y)\}$, we have

$$u_{xx} = f_{\xi\xi}\phi_x^2 + 2f_{\xi\eta}\phi_x\psi_x + f_{\eta\eta}\psi_x^2 + f_\xi\phi_{xx} + f_\eta\psi_{xx},$$

$$u_{xx} = f_{\xi\xi}\phi_x\phi_y + f_{\xi\eta}\left(\phi_x\psi_y + \phi_y\psi_x\right) + f_{\eta\eta}\psi_x\psi_y + f_\xi\phi_{xy} + f_\eta\psi_{xy},$$

$$u_{yy} = f_{\xi\xi}\phi_y^2 + 2f_{\xi\eta}\phi_y\psi_y + f_{\eta\eta}\psi_y^2 + f_\xi\phi_{yy} + f_\eta\psi_{yy}.$$

Change of the Independent Variables

A particularly important type of compound function occurs in the process of changing the independent variables. For example, let $u = f(\xi,\eta)$ be a function of ξ and η, which we interpret as rectangular co-ordinates in the $\xi\eta$-plane. If we rotate the axes in the $\xi\eta$-planes through an angle θ, we obtain a new system of co-ordinates x,y, related to the co-ordinates (ξ,η) by the equations:

$$\xi = x\cos\theta - yx\sin\theta,\ \eta = x\sin\theta + y\cos\theta, \text{ or}$$

$$x = \xi\cos\theta + \eta\sin\theta,\ y = x\sin\theta + y\cos\theta, \text{ or}$$

The function $u = f(\xi,\eta)$ can then be expressed as a function of the new variables x, y:

$$u = f(\xi,\ \eta) = f(x,\ y).$$

The chain rule yields then immediately:

$$u_x = u_\xi\cos\theta + u_\eta\sin\theta,\ u_v = -u_\xi\sin\theta + u_\eta\cos\theta$$

Thus, the partial derivatives are transformed by the same formulae as the independent variables. This is true for rotation of the axes in space also. Another important type of change of co-ordinates is the change from rectangular co-ordinates x, y to polar co-ordinates r, θ.

This is done by means of the equations,

$$x = r\cos\theta,$$
$$y = r\sin\theta,$$
$$r = \sqrt{\left(x^2 + y^2\right)}$$
$$\theta = \arctan\frac{y}{x}$$

We then find that for an arbitrary function $u = f(x,x)$ with continuous partial derivatives of the first order we have

$$u = f(x, y) = f(r \cos \theta, r \sin \theta) = F(r, \theta),$$

$$u_x = u_r r_x + u_\theta \theta_x = u_r \frac{x}{r} - u_\theta \frac{y}{r^2} = u_r \cos \theta - u_\theta \frac{\sin\theta}{r},$$

$$u_y = u_r r_y + u_\theta \theta_y = u_r \frac{y}{r} - u_\theta \frac{x}{r^2} = u_r \sin \theta - u_\theta \frac{\cos\theta}{r}.$$

Whence we obtain the often useful equation

$$u_x^2 + u_r^2 = u_r^2 + \frac{1}{r^2} u_\theta^2$$

In general, let us consider a pair of functions $\xi = \phi(x, y)$, $\eta = \psi(x, y)$ which are continuous and have continuous derivatives in a region R of the xy-plane. These equations assign to each point (x, y) in R a point $\xi = \phi(x, y)$, $\eta = \psi(x, y)$ in the $\xi\eta$-plane. As (x, y) ranges over R, the corresponding point (ξ, η) will range over some set of values S in the $\xi\eta$-plane. Naturally, it is possible that several distinct points (x, y) will give the same values for (ξ, η), so that there corre-spond to several points (x, y) only one point (ξ, η).

We shall assume that this is not s, but instead that there corresponds to one point $Q(\xi, \eta)$ in S exactly one point $P(x,y)$ in R. We may therefore look at the correspondence from either point of view - saying that Q corresponds to P or that P corresponds to Q.

The latter point of view can be expressed as follows: There corresponds to each point (ξ, η) in S one x and one y, namely, the co-ordinates of P, or, in, equations, there are two functions $x = g(\xi, \eta)$, $y = h(\xi, \eta)$, defined in S, which represent the corresponding inverse to $\xi = \phi(x,y)$, $\eta = \psi(x,y)$.

It happens often that the functions $g(\xi, \eta)$, $h(\xi, \eta)$ are by no means easy to calculate, even when they do exist, whence we shall now discover how to obtain the partial derivatives g_ξ, g_η, h_ξ, h_η directly from the partial derivatives ϕ_x, ϕ_y, ψ_x, ψ_y without at all calculating g and h themselves.

For this purpose, we observe that, if we choose any point $Q(\xi, \eta)$, find the corresponding point $P\{g(\xi, \eta), h(\xi, \eta)\}$ in R and then find the point in S corresponding to P, which is $\phi\{g(\xi,\eta),h(\xi,\eta)\}$, $\psi\{g(\xi,\eta),h(\xi,\eta)\}$, we have simply returned to the point Q. In other words, the equations $\xi = \phi\{g(\xi,\eta),h(\xi,\eta)\}$, $\eta = \psi\{g(\xi,\eta),h(\xi,\eta)\}$ are identities in ξ and η. We now differentiate both sides of both equations with respect to ξ and η and find

$$1= \phi_x g_\xi + \phi_y h_\xi,\ 0 = \phi_x g_\eta + \phi_y h_\eta,$$

$$0= \psi_x g_\xi + \psi_y h_\xi,\ 1 = \psi_x g_\eta + \psi_y h_\eta,$$

If an equation expresses an identical relationship, differentiation with respect to any independent variable in it yields an identity, as follows immediately from the definition.

By solving these equations, we find that

$$g_\xi = \frac{\psi_y}{D}, g_n = -\frac{\phi_y}{D}, h_\xi = -\frac{\psi_x}{D}, h_\eta = \frac{\phi_x}{D}$$

or

$$g_\xi = \frac{\eta_y}{D}, x_n = -\frac{\xi_y}{D}, y_\xi = -\frac{\eta_x}{D}, y_\eta = \frac{\xi_x}{D}$$

where by D is the determinant

$$D = \xi x \eta_v - \xi_y \eta_x = \begin{vmatrix} \frac{\partial \xi}{\partial x} & \frac{\partial \xi}{\partial y} \\ \frac{\partial \eta}{\partial x} & \frac{\partial \eta}{\partial y} \end{vmatrix}$$

which we assume to be non-zero. This determinant D, called the functional determinant or Jacobian of (ξ,η) with respect to (x,y), occurs so frequently that one uses frequently the special symbol

$$D = \frac{\partial(\xi,\eta)}{\partial(x,y)}$$

Implicit Functions

In our study of functions of several variables, we have as yet had no analogue to the inverse function. We can regard the inverse function of $y = f(x)$ to be the function obtained by solving the equation $y - f(x) = 0$ for x. In this section, we shall seek more generally to solve equations $F(x, y) = 0$ for x or y and to discuss functions of several variables in a corresponding way.

Even in elementary analytical geometry, curves are frequently represented not by equations $y = f(x)$ or $x = \phi(y)$, but by an equation involving x and y in the form $F(x, y) = 0$. For example, we have the circle $x + y - 1 = 0$, the ellipse $x/a + y/b - 1 = 0$ and. the lemniscate $(x + y) - 2a(x - y) = 0$. In order to obtain y as a function of x or x as a function of y, we must solve the equation for y or for x. We then say that the function $y = f(x)$ or $x = \phi(y)$ found in this way is defined implicitly by the equation $F(x, y) = 0$, and that the solution of this equation yields the function explicitly.

In the examples cited and in many others, the solution can be carried out and the solutions stated explicitly in terms of the elementary functions. In other cases, the solution can be obtained in terms of an infinite series or other limiting processes, i.e., we can approximate to the solution $y = f(x)$ or $x = \phi(y)$ as closely as we please. However, for many purposes, it is more convenient to base the discussion on the implicit definition $F(x, y) = 0$, instead of resorting to an exact or approximate solution of the equation.

The idea that every function $F(x, y)$ yields a function $y = f(x)$ or $x = \phi(y)$ given implicitly by means of the equation $F(x, y) = 0$ is erroneous. On the contrary, it is easy to give examples of functions $F(x, y)$ which, when equated to zero, permit no solution in terms of functions of one variable. Thus, for example, the equation $x + y = 0$ is satisfied by the single pair of values $x = 0$, $y = 0$ only, while the equation $x + y + 1 = 0$ is not satisfied by any real values. It is therefore necessary to investigate this matter more closely, in order to find out whether an equation $F(x, y) = 0$ can actually be solved.

Geometrical Interpretation of Implicit Functions

In order to discuss this problem geometrically, we represent the function $u = F(x, y)$ by a surface in three-dimensional space.

Finding values (x, y) which satisfy the equation $F(x, y) = 0$ is the same thing as finding values (x, y) which satisfy two equations $F(x, y) = u$, $u = 0$; in other words, we wish to find the intersection of the surface $u = F(x, y)$ and the plane $u = 0$ - the xy-plane. We then suppose that we have a definite point (x_0,y_0) which satisfies the equation $F(x_0, y_0) = 0$), i.e., at (x_0, y_0), the surface $u = F(x, y)$ has a point in common with the plane $u = 0$. (If there does not exist such a point, there is no intersection and the equation $F(x, y) = 0$ cannot be solved.)

If the tangent plane to the surface $u = F(x, y)$ at the point (x_0,y_0) is not horizontal, it cuts the plane $u = 0$ in a single straight line. Intuition then tells us that the surface $u = F(x, y)$, lying near the tangent plane, likewise cuts the plane $u = 0$ in a single well defined curve.

How far this curve extends does not at present concern us. The tangent plane will be horizontal, if both the curves $u = F(x_0, y)$ and $u = F(x, y_0)$ have horizontal tangent lines at (x_0, y_0), i.e., if $F_x(x_0, y_0) = 0$ and $F_y(x_0, y_0)) = 0$. Thus, if either $F_x(x_0, y_0) \neq 0$ or $F_y(x_0, y_0) \neq 0$, the tangent plane is not horizontal, and, as we ave just seen, we may expect that a solution in the form $y = f(x)$ or $x = \phi(y)$ hwill exist.

On the other hand, if both $F_x(x_0, y_0)$ and $F_y(x_0, y_0)$ have the value 0, we readily see that there is no guarantee that a solution is possible.

For example, for $F = 1 - \sqrt{\left(1 - x^2 - y^2\right)}$, the corresponding spherical surface is $u = 1 - \sqrt{\left(1 - x^2 - y^2\right)}$, has the point (0, 0) in connection with the xy-plane. The partial derivatives $F_x(0,0)$ and $F_y(0,0)$ are both zero; we find that no point other than (0,0) satisfies the equation $F = 0$. For the function $F(x,y)=xy$, we find that $F(0, 0) = 0$, while $F_x(0, 0) = F_y(0, 0) = 0$.

Here all the points on the x-axis and all the points on the y-axis satisfy the equation $F(x,y)=0$; in the neighbourhood of the origin, we have no unique solution $x=\phi(y)$ or $y=f(x)$. Thus, we see that when $F_x(x_0,y_0) = F_y(x_0,y_0) = 0$, we cannot be sure that a solution exists.

Accordingly, we return to the case in which one of the partial derivatives—say, $F_y(x_0,y_0)$ to be specific—is not zero, the graphical suggestion that a smooth surface should be cut by a non-tangent plane in a smooth curve leads us to expect that one has the theorem:

If the function $F(x,y)$ has continuous derivatives $F_{\dot{x}}$ and F_y, and if at the point (x_0,y_0) holds the equation $F(x_0,y_0) = 0$, while $F_y(x_0,y_0)$ is not zero, then we can mark off about the point (x_0,y_0) a rectangle $x_1 < x < x_2$, $y_1 < y < y_2$ such that for every x in the interval $x_1 < x < x_2$ the equation $F(x, y) = 0$ determines just value $y = f(x)$, lying in the interval $y_1 < y < y_2$. This function $y = f(x)$ satisfies the equation $y = f(x_0)$, and the equation $F\{x, f(x)\} = 0$ is satisfied for every x in the interval. Moreover, the function $y = f(x)$ is continuous and has a continuous derivative. Assuming it to be true, we can add the following: The derivative of the function $y = f(x)$ is given by the equation,

$$y' = f'(x) = -\frac{F_x}{F_y}$$

This follows immediately by using the chain rule. In fact,

$$\frac{d}{dx}F\{x, f(x)\} = F_x\frac{dx}{dx} + F_y\frac{dx}{dx} = F_x + F_y f'$$

However, since $F\{x, f(x)\}$ identically zero, its derivative is also zero, whence $F_x + F_y f' = 0$, and the formula is established.

If we regard the right hand side of the formula as a compound function of x and differentiate according to the chain rule, replacing y' by $-F_x/F_y$, we have

$$y'' = -\frac{F_y\left(F_{xx} + F_{yx}y'\right) - F_x\left(F_{xy} + F_{yy}y'\right)}{F_y^2}$$

$$= -\frac{F_{xx}F_y^2 - 2F_{xy}F_xF_y + F_{yy}F_x^2}{F_y^3}$$

Continuing the process, we may calculate y^{iii}, ^{iv}, etc.

By using this formula, we can usually find the derivative of a function given in implicit form much more easily than by solving first and then differentiating.

For example, for the circle,

$$F(x,y) = x^2 + y^2 - 1 = 0$$

we have,

$$y' = -\frac{F_x}{F_y} = -\frac{x}{y}$$

This is easily verified. In fact, solving the equation of the circle for y, we obtain two solutions, namely,

$$y = \sqrt{\left(1 - x^2\right)} \text{ and } y = -\sqrt{\left(1 - x^2\right)}$$

giving the upper and lower semi-circles, respectively. For the upper part,

$$y' = \frac{-x}{\sqrt{\left(1-x^2\right)}} \text{, for the lower } y' = \frac{+x}{\sqrt{\left(1-x^2\right)}}$$

so that in either case,

$$y' = -\frac{x}{y}$$

Another example is $F(x,y) = e^{x+y} + y - x = 0$. We find $F_x = 0$, while $F_y = 2$. Thus, the equation has a solution $y = f(x)$; but its actual explicit calculation is not simple. Nevertheless, we have

$$y' = -\frac{F_x}{F_y} = -\frac{e^{x+y} - 1}{e^{x+y} + 1}$$

In order that the function f(x) may have a maximum or minimum, we must have $y' = 0$, i.e., $e^{x+y} - 1 = 0$ whence $y = -x$. Substitution of $y = -x$ into the equation $F(x, y) = 0$ yields $1 - 2x = 0$, whence $x = 0, y = -0$. If we calculate $f''(x)$ for $x = 0$, we find it to be negative so is the maximum of y.

8

Differentiating Functions

Differentiation is the process of finding the rate of change of a function. We have proven that if *f* is a variable dependent on an independent variable *x*, such that $f(x) = x^n$ then $\frac{df}{dx} = n.x^{n-1}$ where n is a positive integer.

The derivative reflects the instantaneous rate of change of the function at any value *x*.

The derivative is also a function of *x* whose value is dependent on *x*. Take a look at the left side of the function, $f(x) = x^n$ By definition the derivative of a dependent variable, *f*, is $\frac{df}{dx}$, which is the instantaneous rate of change of *f* with respect to *x* at any condition *x*. The right side of the function, x^n, represents the independent variable whose derivative is $n.x^{n-1}$. When differentiating a function of the form $f(x) = x^n$, the derivative of the dependent variable is, $\frac{df}{dx}$, and the derivative of the independent variable is $n.x^{n-1}$.

Thus differentiating a function results in a new function of *x*, where $\frac{df}{dx} = n.x^{n-1}$. The derivative is called $f'(x)$, read "*f* prime of *x*", and it represents the derivative of a function of *x* with respect to the independent variable, *x*.. If $f'(x) = x^n$, then:

$$\frac{df}{dx} = f'(x) = n.x^{n-1}$$

$f'(x)$ gives the instantaneous rate of change of *f(x)* as a function of any value, *x*. Remember that the rate of change of a function other than a line is not constant. Its value changes as *x* changes.

If f(x) were equal to a constant multiplied by a function of x such as:

$$f(x) = c.x^{n_1}$$
$$f(x) = c.A(x)$$

where $A(x) = x^{n_1}$

The derivative of *f*(*x*) would be:

$$\frac{df}{dx} = \lim_{\Delta x \to 0} \frac{f(x+\Delta x) - f(x)}{\Delta x}$$
$$= c.\frac{A(x+\Delta x) - A(x)}{\Delta x}$$
$$\frac{df}{dx} = f'(x) = c.A'(x)$$

Thus the derivative of *f(x)* with respect to *x*, is the constant multiplied by the derivative of the function of *x*, *A(x)*.

DIFFERENTIATING SUMS OF FUNCTIONS

$f(x)$ equals $x^{n_1} + x^{n_2}$ which is a sum of two functions of *x*, *A*(*x*) and *b*(*x*) where $A(x) = x^{n1}$ and $B(x) = x^{n2}$ Therefore, if $f(x) = A(x) + B(x)$ What would $f'(x)$ be?

The answer is that the derivative is the sum of the derivatives of the two functions To prove this let us return to the definition of the derivative.

We can express a small change in *f*, Δf, equal to $f(x+\Delta x) - f(x)$.

Therefore:

$$\Delta f = f(x+\Delta x) - f(x) = [A(x+\Delta x) + B(x+\Delta x)]$$
$$-[A(x) + B(x)]$$

Δx and taking the limit as Δx goes to zero gives us the instantaneous rate of change of *f* with respect to *x*.

$$\frac{df}{dx} = \lim_{\Delta x \to 0} \frac{f(x+\Delta x) - f(x)}{\Delta x}$$

$$= \frac{[A(x+\Delta x) + B(x+\Delta x)] - [A(x) + B(x)]}{\Delta x}$$

Combing the $A(x)$ and $B(x)$ terms together simplifies the above expression to:

$$\frac{df}{dx} = \lim_{\Delta x \to 0} \frac{f(x+\Delta x) - f(x)}{\Delta x}$$

$$= \lim_{\Delta x \to 0} \frac{[A(x+\Delta x) - A(x)]}{\Delta x} + \frac{[B(x+\Delta x) - B(x)]}{\Delta x}$$

Which reduces to:

$$\frac{df}{dx} = f'(x) = A'(x) + B'(x)$$

Therefore if $f(x)$ is a sum of two functions of x, then its derivative *with respect to* x is the sum of the derivatives of the functions with respect to x. Thus:

$$f(x) = x^{n_1} + x^{n_2}$$

$$\frac{df}{dx} = f'(x) = n_1 \cdot x^{n_1 - 1} + n_2 \cdot x^{n_2 - 1}$$

Similarly if $f(x)$ is defined in terms of a difference among some functions of x, then $f'(x)$ is the sum of the difference among the derivatives of the functions.

$$f(x) = x^{n_1} - x^{n_2} - x^{n_3} \ldots\ldots\ldots$$

$$\frac{df}{dx} = f'(x) = n_1 \cdot x^{n_1 - 1} - n_2 \cdot x^{n_2 - 1} - n_3 \cdot x^{n3-1} \ldots\ldots\ldots$$

Differentiating Products of Functions

Consider the following function:

$$f(x) = x^2 \cdot (x+1)^3$$

If we let , then $f(x)$ can be expressed as the product of the two function $A(x)$ and $B(x)$ such that:

$$f(x) = A(x) \cdot B(x)$$

We can differentiate products of functions by using the definition of the derivative. A small change in *f* can be written as:

$$\Delta f = f(x+\Delta x) - f(x) = A(x+\Delta x)B(x+\Delta x) - A(x)B(x)$$

Next, divide by Δx to calculate the rate of change of *f* with respect to *f*:

$$\frac{\Delta f}{\Delta x} = \frac{f(x+\Delta x) - f(x)}{\Delta x}$$

$$= \frac{A(x+\Delta x)B(x+\Delta x) - A(x)B(x)}{\Delta x}$$

Taking the limit as Δx

goes to zero gives us the instantaneous rate of change of *f* with respect to *x*, or the derivative of *f*(*x*).

$$\frac{df}{dx} = \lim_{\Delta x \to 0} \frac{f(x+\Delta x) - f(x)}{\Delta x}$$

$$= \frac{A(x+\Delta x)B(x+\Delta x) - A(x)B(x)}{\Delta x}$$

From the definition of the derivative we know that:

$$\lim_{\Delta x \to 0} \frac{f(x+\Delta x) - f(x)}{\Delta x} = f'(x)$$

Multiplying both sides by this infinitely small Δx

$$f(x+\Delta x) - f(x) = f'(x) \cdot \Delta x$$

$$f(x+\Delta x) = f(x) + f'(x).\Delta x$$

Since both *A*(*x*) and *B*(*x*) are functions of *x*, then $A(x+\Delta x)$ and $B(x+\Delta x)$ can be substituted with $A(x) + A'(x) \cdot \Delta x$ and $B(x) + B'(x) \cdot \Delta x$ respectively. Note that this substitution only holds true for Δx going to zero. We now have:

$$f'(x) = \lim_{\Delta x \to 0} \frac{[(A(x) + A'(x) \cdot \Delta x)(B(x) + B'(x) \cdot \Delta x)]}{\Delta x}$$

$$\lim_{\Delta x \to 0} \frac{-[A(x)B(x)]}{\Delta x}$$

Expanding the numerator:

$$f'(x) = \lim_{\Delta x \to 0} \frac{[A(x)B(x) + A(x)B'(x)\Delta x + A'(x)\Delta x B(x)}{\Delta x}$$

$$\lim_{\Delta x \to 0} \frac{+A'(x)B'(x)\Delta x^2] - [A(x)B(x)]}{\Delta x}$$

Canceling terms and dividing through by Δx reduces it to:

$$f'(x) = \lim_{\Delta x \to 0} A(x)B'(x) + B(x)A'(x) + A'(x)B'(x)\Delta x$$

$$\therefore f'(x) = A(x)B'(x) + B(x)A'(x)$$

Thus the derivative of a function *f*(*x*) that is a product of two functions of *x*, is simply the product of the first function and the derivative of the second function plus the product of the second function and the derivative of the first function.

$$f(x) = A(x) \cdot B(x)$$

$$f'(x) = A(x)B'(x) + B(x)A'(x)$$

Differentiating Functions of any Power N

We have proven that if $f(x) = x^n$ then $f'(x) = n \cdot x^{n-1}$ for *n* equal to a positive integer, i.e. 1, 2, 3, 4 etc. What if *n* were a fraction such that $f(x) = x^{\frac{1}{3}}$.

As we shall prove, the derivative of any function of *x* of the form $f(x) = x^n$ is always $f'(x) = n \cdot x^{n-1}$ where *n* is any real number, positive, negative, or fraction.

This is no coincidence but is because of the way exponents are defined as a continuous operation for any *n*.

Let us consider the first case of $f(x) = x^{\frac{1}{n}}$, where n is any positive integer and 1/n is a fraction.

$$f(x) = x^{\frac{1}{n}}$$

$$f = x^{\frac{1}{n}}$$

Raising both sides to the *n'th* power:

$$f^n = x$$

$$x = f^n$$

Differentiating *x*, with respect to *f* yields:

$$f^n = x$$

$$x = f^n$$

Taking the reciprocal of the function:

$$\frac{dx}{df} = \frac{1}{n \cdot f^{n-1}} = \frac{1}{n} \cdot \frac{1}{f^{(n-1)}}$$

From the definition of the function, we know that $f = x^{\frac{1}{n}}$. Making this substitution:

$$\frac{df}{dx} = \frac{1}{n} \cdot \frac{1}{x^{\frac{1}{n} \cdot (n-1)}}$$

$$\frac{df}{dx} = \frac{1}{n} \cdot \frac{1}{x^{\left(1-\frac{1}{n}\right)}}$$

$$\frac{df}{dx} = \frac{1}{n} \cdot x^{-\left[\left(1-\frac{1}{n}\right)\right]}$$

$$\therefore \quad \frac{df}{dx} = f'(x) = \frac{1}{n} \cdot x^{\frac{1}{n}-1}$$

This concludes the proof that if $f(x) = x^n$ then $f'(x) = n \cdot x^{n-1}$, for any positive *n*, integer or fraction. We will prove it also holds true for n as a negative number.

We can use the product rule to prove that the derivative of $f(x) = x^n$ is $f'(x) = n \cdot x^{n-1}$ for all n, negative real numbers also. Consider the function:

$$f(x) = x^{-n}$$

$$f = \frac{1}{x^n}$$

Multiplying both sides by x^n yields:

$$f.x^n = 1$$

$$f(x) \cdot x^n = 1$$

$$f(x) \cdot A(x) = B(x)$$

where $A(x) = x^n$ and $B(x) = 1$

The left side of the equation represents a product of two functions, f and x^n, and the right side is a constant function, $B(x) = c = 1$.

Since both functions are equal to each then their derivatives must be equal. Using the product rule to differentiate the right side with respect to x results in:

$$\frac{d[f(x) \cdot A(x)]}{dx} = f(x) \cdot A'(x) + A(x) \cdot f'(x)$$

Similarly, differentiating the right side with respect to x yields:

$$B(x) = 1 \cdot x^0$$

$$B'(x) = 0$$

Setting the derivatives of the left and right side equal to each other:

$$f(x) \cdot A'(x) + A(x) \cdot f'(x) = 0$$

Remember that:

$$f(x) = x^{-n} = \frac{1}{x^n}$$

$$A(x) = x^n$$

$$A'(x) = n \cdot x^{n-1}$$

Making the required substitutions:

$$\left(\frac{1}{x^n}\right) \cdot nx^{n-1} + x^n.f'(x) = 0$$

$$\frac{n}{x} + x^n \cdot f'(x) = 0$$

We can now solve for $f'(x)$

$$f'(x) = -\left(\frac{n}{x}\right) \cdot \frac{1}{x^n}$$

$$f'(x) = -n \cdot \frac{1}{x^{1+n}}$$

$$f'(x) = -n \cdot x^{-(1+n)}$$

$$f'(x) = -n \cdot x^{-n+1}$$

Therefore if $f(x) = x^{-n}$ then $f'(x) = -n \cdot x^{-n+1}$, where $-n$ is a negative integer.

We have proven that if $f(x) = x^n$ is $f'(x) = n \cdot x^{n-1}$ for positive real numbers and negative integers. What remains is to prove it true for negative fractional powers as well. To do this let,

$$f(x) = x^{-\frac{1}{n}}$$

$$f(x) = \frac{1}{x^{\frac{1}{n}}}$$

$f(x) = x^{\frac{1}{n}}$. Differentiating both sides with respect to *x* yields:

$$f(x) \cdot \frac{1}{n} . x^{\frac{1}{n}-1} + x^{\frac{1}{n}} \cdot f'(x) = 0$$

Substituting known values for *f(x)* and solving $f'(x)$ for gives us:

$$f'(x) = -\frac{1}{n} \cdot x^{-\frac{1}{n}-1}$$

Thus if $f(x) = x^{-n}$ then $f'(x) = -n \cdot x^{-n+1}$ for any *n*, positive or negative real number.

This is true because of the way exponents are defined as a uniform operation for any *n*.

The purpose of going through all the proofs for the different cases of *n* was to give you a better understanding of how to differentiate functions of *x* with respect to *x*.

DIFFERENTIATING FUNCTIONS OF FUNCTIONS

The last technique of differentiation is for differentiating functions of functions of x or functions of the form, $f(g(x)) = g(x)^n$.

For example consider the function: $f(x) = (x^2+1)^{\frac{1}{2}}$

The derivative is not $f'(x) = \frac{1}{2}(x^2+1)^{\frac{1}{2}-1}$ because we also need to take into consideration the inside function of x, x^2+1. We can replace x^2+1 with $g(x)$ and get:

$$f(g(x)) = g(x)^{\frac{1}{2}}$$

$$f(g) = g^{\frac{1}{2}}$$

To find the derivative of $f(g(x))$ with respect to x, we first need to find the derivative of f with respect to g. From the definition of the derivative:

$$\frac{df}{dg} = f'(g) = \lim_{\Delta g \to 0} \frac{f(g+\Delta g) - f(g)}{\Delta g}$$

Next we find the derivative of $g(x)$, the inside function, with respect to x.

$$\frac{dg}{dx} = g'(x) = \lim_{\Delta x \to 0} \frac{g(x+\Delta x) - g(x)}{\Delta x}$$

Now we multiply the two derivatives $f'(g)$ and $g'(x)$ to get df/dx:

$$f'(g) \cdot g'(x) = \frac{df}{dg} \cdot \frac{dg}{dx}$$

$$= \lim_{\Delta x \to 0} \frac{f(g+\Delta g) - f(g)}{\Delta x} \cdot \frac{g(x+\Delta x) - g(x)}{\Delta x}$$

$$\Delta g = g(x+\Delta x) - g(x) \quad f'(g)g'(x)$$

$$= \frac{df}{dg} \cdot \frac{dg}{dx} = \lim_{\Delta x \to 0} \frac{f(g+\Delta g) - f(g)}{g(x+\Delta x) - g(x)} \cdot \frac{g(x+\Delta x) - g(x)}{\Delta x}$$

Δx goes to zero, $\dfrac{g(x+\Delta x)-g(x)}{\Delta x} g'(x)$ or:

$$\frac{dg}{dx}=g'(x)$$

Similarly, Δg also goes to zero as Δx goes to zero. Multiplying both sides by *dx*.

$$dg = g'(x)\cdot dx$$

Thus as *dx* goes to zero;

$$dg = c.0 = 0$$

As *dx* goes to zero, $\dfrac{f(g+\Delta g)-f(g)}{\Delta g}$ becomes $f'(g)$.

At the same time the in the denominator of,

$$\frac{f(g+\Delta g)-f(g)}{\Delta g}$$

cancels out with Δg the in the numerator of,

$$\frac{g(x+\Delta x)-g(x)}{\Delta x},$$

since they are both the equivalent. To conclude

$$\frac{df}{dg}\cdot\frac{dg}{dx}=\frac{df}{dx}=f'(g)g'(x)$$

Thus the derivative of a function of a function of x with respect to x, is the derivative of the outer function with respect to the inner function, multiplied by the derivative of the inner function with respect to *x*. Some examples will show how this is done.

USING THE FIRST DERIVATIVE

From the definition of the derivative, $f'(x)$ is the derivative of a function $f(x)$. The derivative of a function tells us how fast f is changing relative to the independent variable, x. Thus the derivative refers specifically the rate of change of the anti-derivative function with respect to *x*. Rate of change is also synonymous with the slope of the tangent line to the graph at a particular point. Therefore, a function and its derivative are closely related and knowing just the derivative can tell us a great deal about the behavior of its anti-derivative.

Since the derivative of a function is derived from the definition of a derivative as:

$$\lim_{\Delta \to 0} \frac{f(x+\Delta x)-f(x)}{\Delta x},$$

we can work with the definition to find the anti-derivative when only the derivative is known. Also recall how differentiation is based on a limiting or subtracting process and dividing; therefore working backwards would tell us that we should be adding and multiplying. We will develop on this later, however, let us first look at how we can use $f'(x)$ to obtain equilibrium points on the graph of the anti-derivative or $f(x)$. Equilibrium points are by definition, points on the graph refer to static situations where the rate of change is zero. Thus a change in the independent variable results in no change in the dependent variable. Equilibrium sate generally occurs when a situation has reached a critical maximum value and then decreases or where a situation has reached a critical minimum value then increases.

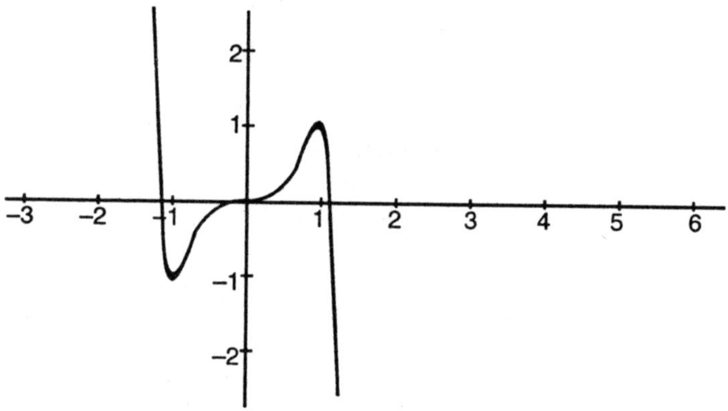

Equilibrium and critical values of a function can refer to different things depending on the phenomena being studied. Therefore we will restrict ourselves to the geometric interpretation of an equilibrium as a point on the graph where the rate of change is zero. Since the rate of change is zero, the tangent to the graph at this point will be a horizontal line. A horizontal tangent tells us that the derivative's value at the equilibrium point is zero. Such situations occur when the graph has reached a maximum or minimum value. Horizontal tangent may also exist, but not necessarily, when the concavity of a graph changes.

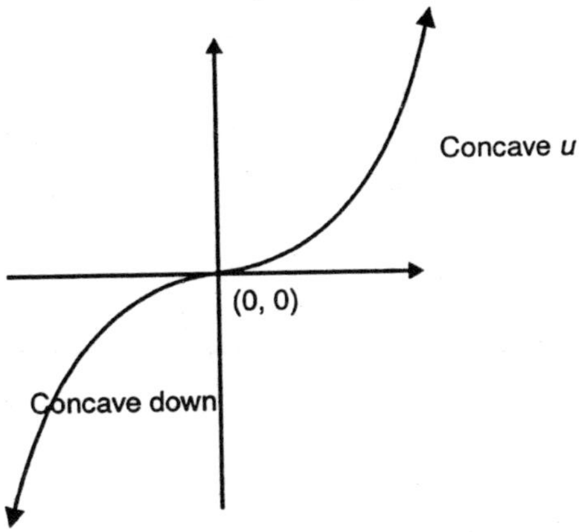

A change in concavity occurs at points on the graph called *inflection* points. As we shall soon study, the derivatives value at an inflection does not have to be zero. Therefore we will restrict our definition of equilibrium points to reflect either maximum or minimum values on the graph.

For example to find equilibrium points for the function, $f(x) = x^3 - x^2$. we first need to differentiate it to get, $f'(x)$. The derivative, $f'(x)$, tells us the instantaneous rate of change of function, $\frac{df}{dx}$, at any point x. Since we want to find points where the rate of change of $f'(x)$ is zero, we need to set equal to zero to find those values of x which the satisfy the equation, $f'(x) = 0$. Doing this for $f(x) = x^3 - x^2$ results in:

$$f'(x) = 3x^2 - 2x$$

$$f'(x) = 0$$

$$3x^2 - 2x = 0$$

$$(x)(3x - 2) = 0$$

$$\therefore x = 0 \text{ and } x = 2/3$$

This tells us that at both $x = 0$ and $x = 2/3$ there exists an equilibrium point, which is confirmed by the graph of the function, $f(x) = x^3 - x^2$

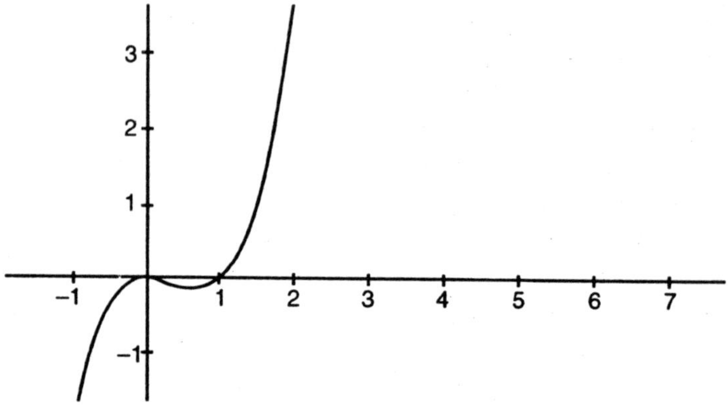

Having found the critical points, how do we classify them as either maximums or minimums? Obviously this can be done by just looking at the graph of $f(x)$, but the purpose of this chapter is to understand how only use the derivative, $f'(x)$ to approximate the behavior of the function $f(x)$. To determine whether our equilibrium points are either maximum or minimums we need to evaluate points left and right of the equilibrium points to determine where the function, $f(x)$, is increasing or decreasing.

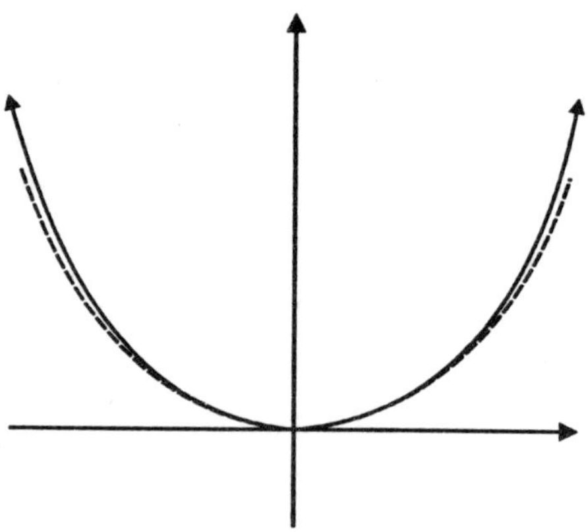

A minimum is defined as the bottom of a U-shaped or concave up graph. If the graph is concave up then the slope or rate of change is positive to the right of an equilibrium point and the function is increasing to the right of that point.

To the left of the equilibrium point, the slope is negative which means the function is decreasing till it reaches the equilibrium point. To better understand this, look at the following graph of a concave up portion of a graph. Notice how the function decreases till it reaches the equilibrium point then rises after passing it.

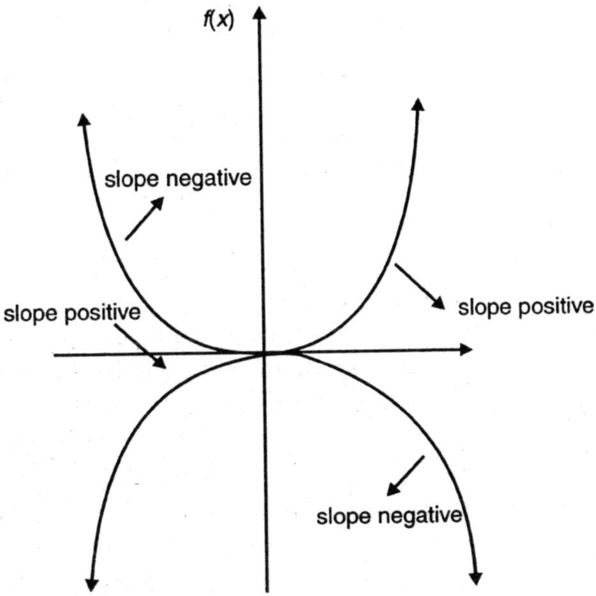

Similarly if the rate of change of $f(x)$ were negative on the right side and positive on the left side of the equilibrium point, then we get an inverted U shaped or concave down graph.

A concave down graph thus reflects a maximum value at the equilibrium point. If the slope is both positive or negative on either side of the equilibrium point then we get an inflection point that represents where the concavity changes from a $\cup$ shape to a $\cap$ shape or vice-versa. The following graph summarizes the above conclusions.

USING THE SECOND DERIVATIVE

The first derivative allows us to define equilibrium points on the graph of a function, $f(x)$. By evaluating points to the left and right of the equilibrium point we can classify these points as either maximums or minimums and thus determine the concavity of the graph. Without having found the equilibrium points it is extremely difficult to determine the behavior of a function over an interval.

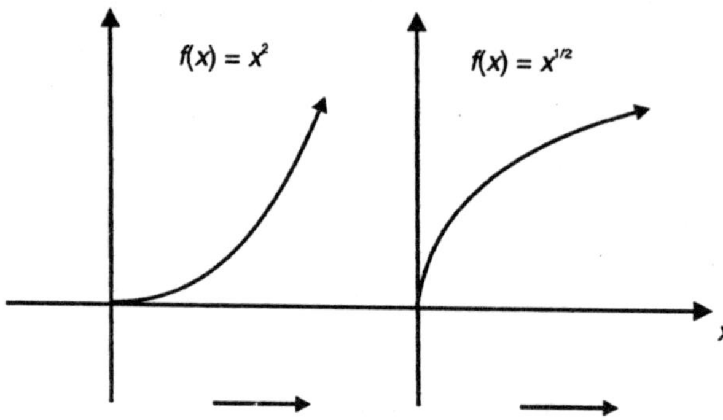

The sign of the first derivative only tells us if a function is increasing or decreasing; however, a function can increase or decrease in two way. For example consider the graphs of the following two different functions.

In both cases the function is increasing and the first derivative is always positive; however each function increases in a different way i.e. one increases concave up and the other increases concave down. Using the first derivative only, we would have to know not only where its positive or negative but also how the first derivative is changing i.e. positive and increasing, negative and increasing etc.

For the first graph $f''(x)$ is positive and increasing thus the graph of $f(x)$ is increasing and concave up. For the second graph, $f''(x)$ is also positive but is decreasing. Thus the graph of $f(x)$ is concave down. The process of looking only at the graph of the first derivative to understand how $f(x)$ behaves is an extremely abstract and difficult one.

To quicken and simplify our work we can use the function's

second derivative to conclude where the graph is concave up or down. This information along with the fact that the derivative is either positive or negative over an interval will be enough to accurately determine a function's behavior. Recall that a property of a concave up part of a graph is that its slope or rate of change is always increasing.

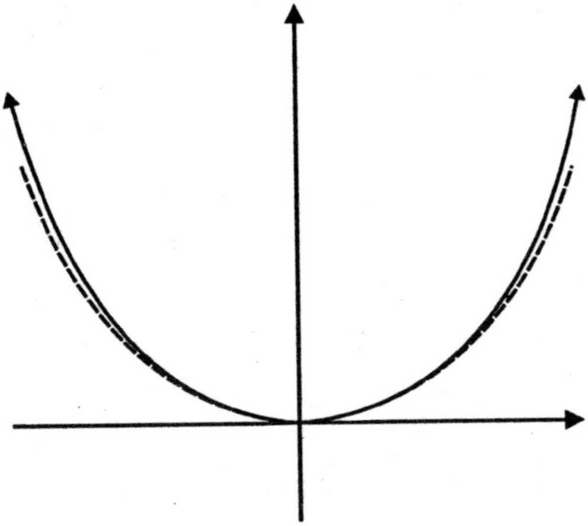

On the left side the slope is negative; however, as x increases the slope gets less and less, –5, –3, –2, till it reaches 0, from where on it increases to 1, 5, 6, etc. We can then conclude that the rate at which the slope is changing must be positive or the graph of the derivative is increasing. Since the derivative's value is constantly increasing, then the rate of change of the derivative, given by $f''(x)$ will be positive. Remember that positive rate of change implies that the function is increasing over that interval, while a negative rate of change implies the function decreases as x increases.

In a concave up graph the derivative is increasing, such that the second derivative over this interval will be positive. Working in reverse we arrive at an important conclusion. If the second derivative is positive over an interval, then the first derivative is increasing, implying that the graph of the original function, $f(x)$ is concave up. This is true because the rate of change of a concave up graph is always increasing.

The reverse is true for concave down graphs. If the second

derivative is negative then the first derivative is decreasing, implying that the original functions graph is concave down over the interval. $f''(x) > 0$ is concaveup $f''(x) < 0$ hence $f(x)$ is concave down. Though all the information concerning the behavior of $f(x)$ can be obtained from studying its derivative, we can quicken and confirm our sketches by looking at the functions second derivative. Without having found any equilibrium points we can accurately determine the behavior of $f(x)$ over an interval by using the signs of both the function's first and second derivative simultaneously. Four possibilities may exist for the signs of the derivatives. Both $f'(x)$ and $f''(x)$ are positive over an interval.

1. Therefore $f(x)$ is increasing and concave up.
2. $f'(x)$ is positive but $f''(x)$ is negative. Thus $f(x)$ is increasing and concave down
3. $f'(x)$ and $f''(x)$ are negative, in which case $f(x)$ is decreasing and concave down.
4. $f'(x)$ is negative but $f''(x)$ is positive, thus $f(x)$ is decreasing and concave up.

SYSTEMATIC USE OF THE DERIVATIVES

We shall now look at a systematic and orderly way of interpreting our knowledge of a function's first and second derivative.

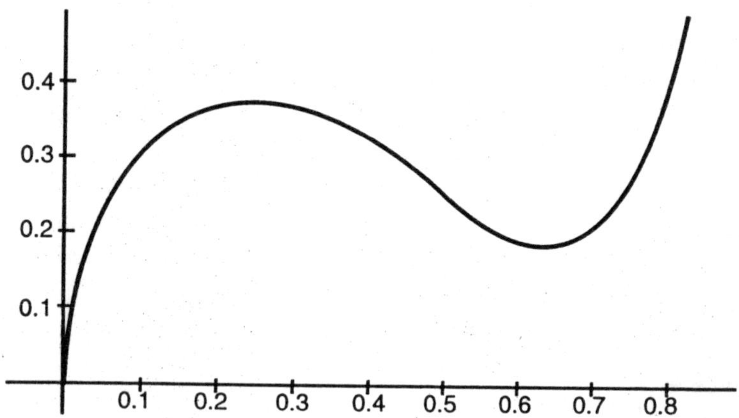

Before continuing let us return to the concept of the inflection point. An inflection point is a point on the graph where the concavity shifts from being concave up to concave down or vice-versa. Since a negative second derivative reflects a concave down graph while a positive second derivative represents a concave up graph then inflection point occur where, $f'(x)$ is equal to zero.

Note that inflection points do not have to exist at equilibrium points. For example in the graph of $f(x)=2x^2-4x^3-\sqrt{x}$, the derivatives value at the inflection point ($x = 0.45$) is not equal to zero. In some rare cases, the value of the derivatives can be of the same sign before and after an equilibrium point. In such cases the graph of the function, has to be determined by carefully looking at both the functions first and second derivative.

To begin drawing $f(x)$, first find maximum and minimum values by setting $f'(x)$ equal to zero, and solving for x. Next find inflection points by setting $f''(x) = 0$ and solving for x. Then plot two number lines for both $f'(x)$ and $f''(x)$ with plus and minus signs to indicate where they are positive and negative.

First look at $f'(x)$ to understand where the graph is increasing or decreasing. Next look at $f''(x)$ to find out how the graph is increasing or decreasing, concave up or down.

The number line for the example is:

$$f(x)=x^3, f'(x)=3x^2; f''(x)=6x$$

f'(x)

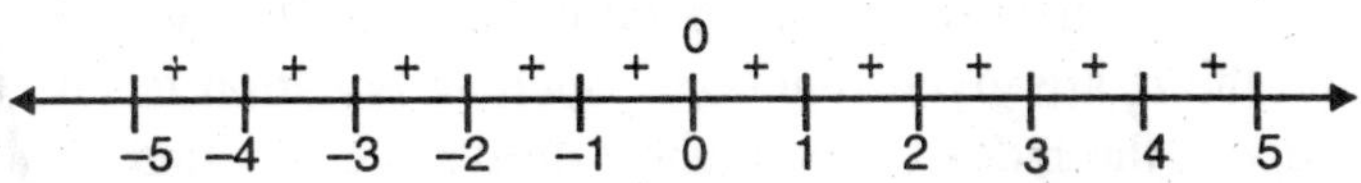

f'(x)

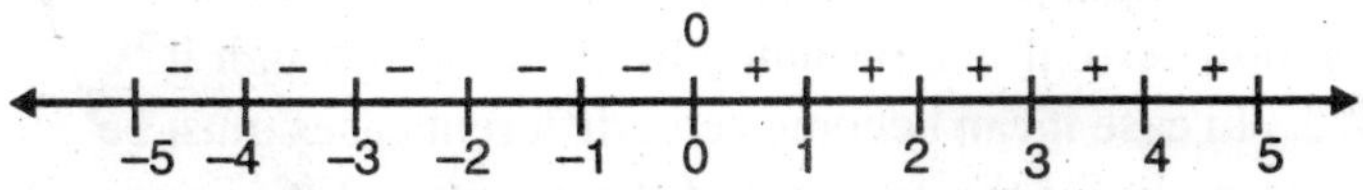

Having drawn this important number line, how do we make interpret it. Remember positive rate of change implies that $f''(x)$ is increasing while the reverse is true for negative rate of change. Similarly a positive second derivative implies that the graph is concave up while negative values represent concave down. Then use equilibrium points to draw the graph through them.

9

Differential Equations

LINEAR DIFFERENTIAL EQUATION

In mathematics, a linear differential equation is a differential equation of the form

$$Ly = f$$

where the differential operator L is a linear operator, y is the unknown function, and the right hand side f is a given function (called the source term). The linearity condition on L rules out operations such as taking the square of the derivative of y; but permits, for example, taking the second derivative of y. Therefore a fairly general form of such an equation would be

$$a_n(x)\, D^n y(x) + a_{n-1}(x)\, D^{n-1} y(x)$$

$$+ \ldots + a_1(x)\, Dy(x) + a_0(x) y(x) = f(x)$$

where D is the differential operator d/dx (i.e. $Dy = y'$, $D^2y = y''$,..), and the a_i are given functions. Such an equation is said to have order n, the index of the highest derivative of y that is involved. (Assuming a possibly existing coefficient a_n of this derivative to be non zero, it is eliminated by dividing through it.

In case it can become zero, different cases must be considered separately for the analysis of the equation.) If y is assumed to be a function of only one variable, one speaks about an ordinary differential equation, else the derivatives and their coefficients must be understood as (contracted) vectors, matrices or tensors of higher rank, and we have a (linear) partial differential equation.

The case where $f = 0$ is called a homogeneous equation and its solutions are called complementary functions. It is particularly important to the solution of the general case, since any complementary function can be added to a solution of the inhomogeneous equation to give another solution (by a method traditionally called *particular integral and complementary function*). When the a_i are numbers, the equation is said to have *constant coefficients*.

The subject of ordinary differential equations encompasses such a large field that you can make a profession of it. There are however a small number of techniques in the subject that you have to know. These are the ones that come up so often in physical systems that you need not only the skills to use them but the intuition about what they will do.

LINEAR CONSTANT-COEFFICIENT

A differential equation such as

$$\left(\frac{d^2x}{dt^2}\right)^3 + t^2x^4 + 1 = 0$$

is difficult to solve, sinceit is non-linear. This means that if I have two solutions $x_1(t)$ and $x_2(t)$ then the sum $x_1 + x_2$ is not a solution; look at all the cross-terms you get if you try to plug the sum in to the equation. Also if you multiply $x_1(t)$ by 2 you no longer have a solution.

An equation such as

$$e^t\frac{d^3x}{dt^3} + t^2\frac{dx}{dt} - x = 0$$

may be a mess to solve, but if you have two solutions, $x_1(t)$ and $x_2(t)$ then the sum $\alpha x_1 + \beta x_2$ is also a solution. This is called a linear, homogeneous equation because of this property. A similar-looking equation,

$$e^t\frac{d^3x}{dt^3} + t^2\frac{dx}{dt} - x = t$$

does not have this property, though it's close. It's called a linear, inhomogeneous equation.

If $x_1(t)$ and $x_2(t)$ are solutions to this, then if you try their sum as a solution you get $2t = t$, and that's no solution, but it misses working only because of the single term on the right. One of the most common sorts of differential equations that you see is an especially simple one to solve. That's part of the reason it's so common.

This is the linear, constant-coefficient, differential equation. If you have a mass tied to the end of a spring and the other end of the spring is fixed, the force applied to the mass by the spring is to a good approximation proportional to the distance that the mass has moved from its equilibrium position.

If the coordinate x is measured from the mass's equilibrium position, the equation $\vec{F} = m\vec{a}$ says

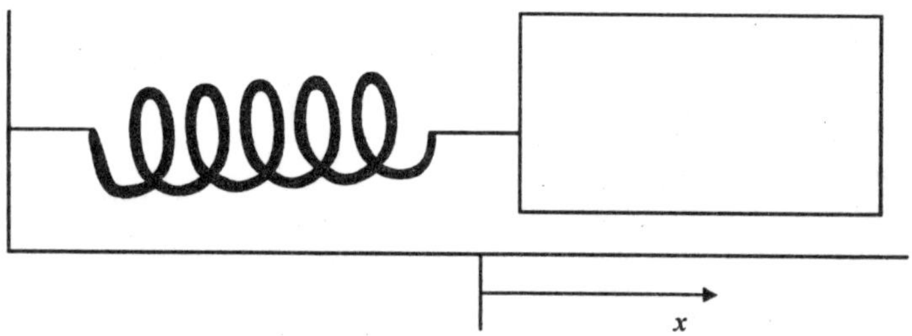

$$m\frac{d^2x}{dt^2} = -kx$$

If there's friction (and there's *always* friction), the force has another term. Now how do you describe friction mathematically? The common model for dry friction is that the magnitude of the force is independent of the magnitude of the mass's velocity and opposite to the direction of the velocity. If you try to write that down in a compact mathematical form you get something like,

$$\vec{F}_{friction} = -\mu_k F_N \frac{\vec{v}}{|\vec{v}|}$$

This is hard to work with. It can be done, but you are going to do something different. Wet friction is easier to handle mathematically because when you lubricate a surface, the friction becomes velocity dependent in a way that is, for low speeds, proportional to the velocity.

$$\vec{F}_{friction} = -b\vec{v}$$

Neither of these two representations is a completely accurate description of the way friction works. That's far more complex than either of these simple models, but these approximations are good enough for many purposes and I'll settle for them. Assume "wet friction" and the differential equation for the motion of *m* is

$$m\frac{d^2x}{dt^2} = -kx - b\frac{dx}{dt}$$

This is a second order, linear, homogeneous, differential equation, which simply means that the highest derivative present is the second, the sum of two solutions is a solution, and a constant multiple of a solution is a solution.

That the coefficients are constants makes this an easy equation to solve. All you have to do is to recall that the derivative of an exponential is an exponential. $de^t/dt = e^t$. If you substitute this exponential for $x(t)$, of course it can't work as a solution; it doesn't even make sense dimensionally. What is *e* to the power of a day?

You have to have something in the exponent to make it dimensionless, $e^{\alpha t}$.

Also, the function *x* is supposed to give you a position, with dimensions of length. Use another constant: $x(t) = Ae^{\alpha t}$. Plug *this* into the differential equation to find

$$mA\alpha^2 e^{\alpha t} + bA\alpha e^{\alpha t} + kAe^{\alpha t} = Ae^{\alpha t}[m\alpha^2 + b\alpha + k] = 0$$

The product of factors is zero, and the only way that a product of two numbers can be zero is if one of the numbers is zero. The exponential never vanishes, and for a non-trivial solution $A \neq 0$, so all that's left is the polynomial in α.

$m\alpha^2 + b\alpha + k = 0$, with solutions $\alpha = \dfrac{-b \pm \sqrt{b^2 - 4km}}{2m}$

The position function is then

$$x(t) = Ae^{\alpha_1 t} + Be^{\alpha_2 t}$$

where A and B are arbitrary constants and α_1 and α_2 are the two roots. Supposed that it is oscillating? It is a harmonic oscillator after all, but the exponentials don't look very oscillatory. If you have a mass on the end of a spring and the entire system is immersed in honey, it won't do much oscillating! Translated into mathematics, this says that if the constant b is too large, there is no oscillation. In the equation for α, if b is large enough the argument of the square root is positive and both α 's are real — no oscillation. Only if b is small enough does the argument of the square root become negative; then you get complex values for the α 's and hence oscillations. Push this to the extreme case where the damping vanishes: $b = 0$.

Then $\alpha_1 = i\sqrt{k/m}$ and $\alpha_2 = -i\sqrt{k/m}$.
Denote $\omega_0 = \sqrt{k/m}$.

$$x(t) = Ae^{i\omega_0 t} + Be^{-i\omega_0 t}$$

You can write this in other forms, see problem 10. To determine the arbitrary constant A and B you need two equations. They come from some additional information about the problem, typically some initial conditions. Take a specific example in which you start from the origin with a kick, $x(0) = 0$ and $\dot{x}(0) = v_0$.

$$x(0) = 0 = A + B, \quad \dot{x}(0) = v0 = i\omega_0 A - \omega_0 B$$

Solve for A and B to get $A = -B = v_0/(2i\omega_0)$. Then

$$x(t) = \frac{v_0}{2i\omega_0}[e^{i\omega_0 t} - e^{-i\omega_0 t}] = \frac{v_0}{\omega_0}\sin\omega_0 t$$

As a check on the algebra, use the first term in the power series expansion of the sine function to see how x behaves for small t. The sine factor is sin $\omega_0 t \approx \omega_0 t$, and then $x(t)$ is approximately $v_0 t$, just as it should be. Also notice that despite all the complex numbers, the final answer is real.

DAMPED OSCILLATOR

If there is damping, but not too much, then the α 's have an imaginary part *and* a negative real part.

$$\alpha = \frac{-b \pm i\sqrt{4km - b^2}}{2m} = -\frac{b}{2m} \pm i\omega', \text{ where}$$

$$\omega' = \sqrt{\frac{k}{m} - \frac{b^2}{4m^2}}$$

This represents a damped oscillation and has frequency a bit lower than the one in the undamped case. Use the same initial conditions as above and you will get similar results (let $(\gamma = b/2m)$

$$x(t) = Ae^{(-\gamma + i\omega')t} + Be^{(-\gamma - i\omega')t}$$

$$x(0) = A + B = 0,\ v_x(0) = (-\gamma + i\omega')A + (-\gamma - i\omega')B = v_0$$

The two equations for the unknowns A and B imply $B = -A$ and

$$2i\omega' A = v_0, \text{ so } x(t) = \frac{v_0}{2i\omega'} e^{-\gamma t}[e^{i\omega' t} - e^{-i\omega' t}] \frac{v_0}{\omega'} e^{-\gamma t} \sin \omega' t$$

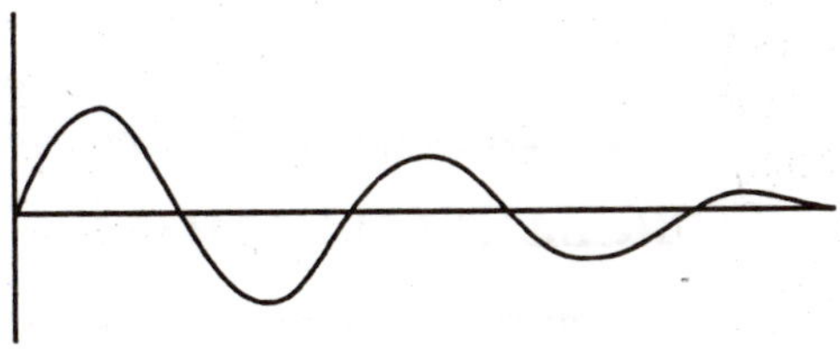

For small values of t, the first terms in the power series expansion of this result are

$$x(t) = \frac{v_0}{\omega'}[1 - \gamma t + \gamma^2 t^2/2 - ...]$$

$$[\omega' t - \omega'^3 t^3/6 + ...] = v_0 t - v_0 \gamma t^2 + ...$$

The first term is what you should expect, as the initial velocity is $v_x = v_0$. The negative sign in the next term says that it doesn't move as far as it would without the damping, but analyze it further. Does it have the right size as well as the right sign? It is $-v_0\gamma t^2 = -v_0(b/2m)t^2$. But that's an acceleration: $a_x t^2/2$. It says that the acceleration just after the motion starts is $a_x = -bv_0/m$. Is that what you should expect? As the motion starts, the mass hasn't gone very far so the spring doesn't yet exert much force. The viscous friction is however $-bv_x$. Set that equal to ma_x and you see that,

$$-v_0\gamma t^2$$

has precisely the right value. In Equation you assumed that the two roots of the quadratic, the two α's, are different. What if they aren't? Then you have only one value of to use in defining the solution $e^{\alpha t}$ in Equation. You now have only one arbitrary constant with which to match two initial conditions. What is the energy for this damped oscillator? The kinetic energy is $mv^2/2$ and the potential energy for the spring is $kx^2/2$. Is the sum constant? No.

$$\frac{d}{dt}\frac{1}{2}(mv^2 + kx^2) = mv\frac{dv}{dt} + kx\frac{dx}{dt}$$

$$= v_x(ma_x + kx) = -bv_x^2 = F_{x,\text{frict}}\, v_x$$

"Force times velocity" is a common expression for power, and this says that the total energy is decreasing according to this formula. The energy decreases exponentially on average.

FORCED OSCILLATIONS

What happen if the equation is inhomogeneous? That is, what if there is a term that doesn't involve x or its derivatives at all. In this harmonic oscillator example, apply an extra external force. Maybe it's a constant; maybe it's an oscillating force; it can be anything you want not involving x.

$$m\frac{d^2x}{dt^2} = -kx - b\frac{dx}{dt} + F_{ext}(t)$$

The key result that you need for this class of equations is very simple to state and not too difficult to implement. It is a procedure for attacking any linear inhomogeneous differential equation and consists of three steps.

1 Temporarily throw out the inhomogeneous term [here $F_{ext}(t)$] and completely solve the resulting homogeneous equation.
In the current case that's what you just got through doing when you solved the equation $md^2x/dt^2 + bdx/dt + kx = 0$.
$[x_{hom}(t)]$
2. Find any one solution to the full inhomog eneous equation. Note that for step one you have to have all the arbitrary constants present; for step two you do not. $[x_{inh}(t)]$
3. Add the results of steps one and two. $[x_{hom}(t) + x_{inh}(t)]$

You 've already done step one. To carry out the next step you'll start with a particular case of the forcing function. If $F_{ext}(t)$ is simple enough, you should be able to guess the answer to step two. If it's a constant, then a constant will work for x. If it's a sine or cosine, then you can guess that a sine or cosine or a combination of the two should work. If it's an exponential, then guess an exponential — remember that the derivative of an exponential is an exponential. If it's the sum of two terms, such as a constant and an exponential, it's easy to verify that you add the results that you get for the two cases separately. If the forcing function is too complicated for you to guess a solution then there's a general method using Green's functions.

Choose a specific example

$$F_{ext}(t) = F_0[1 - e^{-\beta t}]$$

This starts at zero and builds up to a final value of F_0. It does it slowly or quickly depending on β.

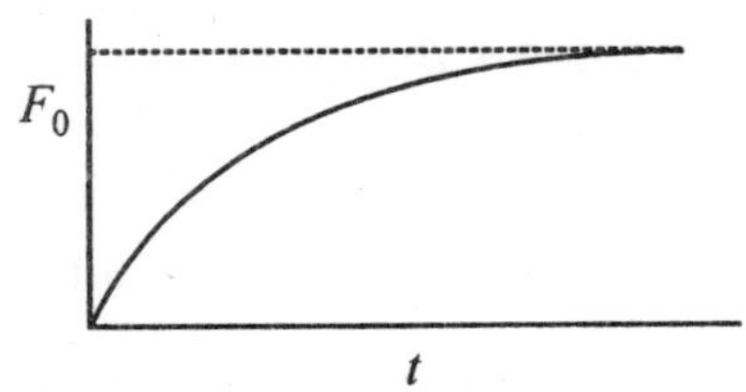

Start with the first term, F_0, for external force in Equation.

Try $x(t) = C$ and plug into that equation to find,

$$kC = F_0$$

This is simple and determines C. Next, use the second term as the forcing function, $-F_0 e^{-\beta t}$. Guess a solution $x(t) = C_0 e^{-\beta t}$ and plug in. The exponential cancels, leaving

$$mC'\beta^2 - bC'\beta + kC' = -F_0 \quad \text{or} \quad C' \frac{-F_0}{m\beta^2 - b\beta + k}$$

The total solution for the inhomogeneous part of the equation is then the sum of these two expressions.

$$x_{inh}(t) = F_0 = \left(\frac{1}{k} - \frac{1}{m\beta^2 - b\beta + k} e^{-\beta t} \right)$$

The homogeneous part of Equation has the solution found in Equation and the total is

$$x(t) = x_{hom}(t) + x_{inh}(t) = x(t)$$

$$= Ae^{\alpha_1 t} + Be^{\alpha_2 t} + F_0 \left(\frac{1}{k} - \frac{1}{m\beta^2 - b\beta + k} e^{-\beta t} \right)$$

There are two arbitrary constants here, and this is what you need because you have to be able to specify the initial position and the initial velocity independently; this is a second order differential equation after all. Take for example the conditions that the initial position is zero and the initial velocity is zero. Everything is at rest until you start applying the external force. This provides two equations for the two unknowns.

$$x(0) = 0 = A + B + F0 \frac{m\beta^2 - b\beta}{k(m\beta^2 - b\beta + k)}$$

$$\dot{x}(0) = 0 = A\alpha_1 + B\alpha_2 + F_0 \frac{\beta}{m\beta^2 - b\beta + k}$$

Now all you have to do is solve the two equations in the two unknowns A and B. I would take the first, multiply it by α_1 and α_2 subtract the second. This gives A. Do the same with 1 instead of α_2 to get B.

The results are,

$$A = \frac{1}{\alpha_1 - \alpha_2} F_0 \frac{\alpha_2(m\beta^2 - b/\beta) - k\beta}{k(m\beta^2 - b\beta + k)}$$

Interchange α_1 and α_2 to get B.

The problem isn't finished until you've analyzed the supposed solution. After all, you may have made some errors in algebra along the way. Also, analyzing the solution is the way you learn how these functions work.

1. Everything in the solution is proportional to F_0 and that's not surprising.
2. It is as an exercise to check the dimensions.
3. A key parameter to vary is β. What should happen if it is either very large or very small? In the former case the exponential function in the force drops to zero quickly so the force jumps from zero to F_0 in a very short time – a step in the limit that $\beta \to 0$.
4. If β is very small the force turns on very gradually and gently, being careful not to disturb the system.

For large β the dominant terms in both numerator and denominator everywhere are the $m\beta^2$ terms. This result is then very nearly,

$$x(t) \approx \frac{F_0}{\alpha_1 - \alpha_2} \frac{(\alpha_2(m/\beta^2))e^{\alpha_1 t} - (\alpha_1(m/\beta^2))e^{\alpha_2 t}}{km/\beta^2}$$

$$+F_0\left(\frac{1}{k} - \frac{1}{(m\beta^2)}e^{-\beta t}\right)$$

$$\approx \frac{F_0}{k(\alpha_1 - \alpha_2)}[(\alpha_2 e^{\alpha_1 t} - \alpha_1 e^{\alpha_2 t}] + F_0\frac{1}{k}$$

Use the notation of Equation and you have,

$$x(t) \approx \frac{F_0}{k(-\gamma + i\omega' - (-\gamma - \omega'))}[((-\gamma - i\omega')e^{(-\gamma + i\omega')t}$$

$$-(-\gamma + i\omega')e^{(-\gamma - i\omega)t}] + F_0\frac{1}{k}$$

$$= \frac{F_0 e^{-\gamma t}}{k(2i\omega')}[-2i\gamma \sin\omega' t - 2i\omega' \cos\omega' t] + F_0 \frac{1}{k}$$

$$= \frac{F_0 e^{-\gamma t}}{k}\left[-\frac{\gamma}{\omega'}\sin\omega' t - \cos\omega' t\right] + F_0 \frac{1}{k}$$

At time $t = 0$ this is still zero even with the approximations. That's comforting, but if it hadn't happened it's not an insurmountable disaster. This is an approximation to the exact answer after all, so it could happen that the initial conditions are obeyed only approximately. The exponential terms have oscillations and damping, so the mass oscillates about its eventual equilibrium position and after a long enough time the oscillations die out and you are left with the equilibrium solution $x = F_0/k$. For small β the β^2 terms in Equation are small compared to the β terms to which they are added or subtracted. The numerators of the terms with $e^{\alpha t}$ are then proportional to β. The denominator of the same terms has a $k - b\beta$ in it. That means that it doesn't go to zero as β goes to zero. The last terms, that came from the inhomogeneous part, don't have any β in the numerator so they don't vanish in this limit. The approximate final result then comes just from the $x_{inh}(t)$ term.

$$x(t) \approx F_0 \frac{1}{k}(1 - e^{-\beta t})$$

It doesn't oscillate at all and just gradually moves from equilibrium to equilibrium as time goes on. It's what you get if you go back to the differential equation and say that the acceleration and the velocity are negligible.

$$m\frac{d^2x}{dt^2}[\approx 0] = -kx - b\frac{dx}{dt}[\approx 0] + f_{ext}(t) \Rightarrow x \approx \frac{1}{k}F_{ext}(t)$$

The spring force nearly balances the external force at all times; this is "quasi-equilibrium."

Bibliography

Anuradha ,H.: *Differential Equations,* Cambridge: Cambridge University Press, 1999.

Aris, R.: *Elementary Illustrations of the Differential and Integral Calculus,* Berkeley: University of California Press, 2005.

Avez ,A.: *An Elementary Treatise on the Differential Calculus,* Princeton: Princeton University Press, 2001.

Caren, L.: *Calculus Collection: A Resource for AP and Beyond,* Providence: CRC-Press, 2001.

Colombeau , J.: *Differential Calculus and Holomorphy: Real and Complex Analysis,* USA: Oxford University Press, 2002.

Greenspan, D.: *Introduction to Partial Differential Equations,* USA: Oxford University Press, 1998.

Guinness, I.: *Landmark Writings in Western Mathematics,* New York: Chapman and Hall, 2000.

Morgan , A.: *Capsule Calculus,* Princeton: Princeton University Press, 2001.

Paul, D.: *Differential Equations with Applications,* Princeton: Princeton University Press, 2004.

Rao, S.: *Differential Calculus,* Berkeley: University of California Press, 2004.

Rhee, H.: *First-Order Partial Differential Equations,* Princeton: Princeton University Press, 2000.

Ritow , I.: *Calculus of Variations: With Applications to Physics and Engineering,* Princeton: Princeton University Press, 2000.

Rodino , L.: *Global Pseudo-Differential Calculus on Euclidean Spaces,* Berkeley: University of California Press, 2000.

Thayse , A.: *Boolean Calculus of Differences,* Cambridge: Cambridge University Press, 2002.

Washington , B.: *Noncommutative Differential Calculus from the Inner Derivation,* USA: Oxford University Press, 2000.

Weinstock , R.: *Circular of Information,* Berkeley: University of California Press, 2002.

William, T.: *Calculus of Variations: The Lagrangian Formalism,* Cambridge: Cambridge University Press, 2000.

Index